**Rapid Assessment Program**

# A biological assessment of the terrestrial ecosystems of the Draw River, Boi-Tano, Tano Nimiri and Krokosua Hills forest reserves, southwestern Ghana

# Bulletin *of* Biological Assessment

# 36

**Editors**
Jennifer McCullough, Jan Decher, and David Guba Kpelle

Center for Applied Biodiversity Science (CABS)

Conservation International - Ghana

The *RAP Bulletin of Biological Assessment* is published by:
Conservation International
Center for Applied Biodiversity Science
1919 M Street NW, Suite 600
Washington, DC 20036

202-912-1000 tel
202-912-1030 fax
www.conservation.org
www.biodiversityscience.org

**RAP *Bulletin of Biological Assessment* Series Editors**
**Terrestrial and AquaRAP:**
Leeanne E. Alonso
Jennifer McCullough
**Marine RAP:**
Sheila A. McKenna
**Design:** Glenda P. Fábregas
**Map:** Mark Denil
**Photography:** Jan Decher, Raffael Ernst, Torben Larsen and Jennifer McCullough

ISBN 1-881173-90-9
©2005 Conservation International

Library of Congress Card Catalog Number 2005932687
DOI: 10.1896/ci.cabs.2005.rap 32.

**Suggested Citation**
McCullough, J., J. Decher and D. Guba Kpelle (eds.). 2005. A biological assessment of the terrestrial ecosystems of the Draw River, Boi-Tano, Tano Nimiri and Krokosua Hills forest reserves, southwestern Ghana. RAP Bulletin of Biological Assessment 36. Conservation International, Washington, DC.

Financial support for this RAP expedition and publication was generously provided by the Critical Ecosystem Partnership Fund (CEPF), Global Conservation Fund (GCF), and the Giuliani Family Foundation.

# Table of Contents

A biological assessment of the terrestrial ecosystems of the Draw River, Boi-Tano,
Tano Nimiri and Krokosua Hills forest reserves, southwestern Ghana

3

# Acknowledgements

The success of this RAP survey was due to the support of many people. The participants thank the staff of CI-Ghana, especially the Country Director, Okyeame Ampadu-Agyei, Bruno Achana, and David Guba Kpelle, for the organization of the logistics in Ghana and for the opportunity to participate in the RAP survey. We also thank the Forestry Commission of Ghana for permitting access to the forest reserves. We owe a debt of gratitude to Isaac Sopelle our cook who catered to our diverse appetites and our drivers, Emmanuel Asane-Dadzie, Samuel Appiah, Aaron Buckman Apietu and Peter Clever Kpiebemane, who took us safely through the difficult terrain of the Western Region of Ghana.

In the Draw River area we received all possible support from the staff of the Forest Services Division, Mohammed Nuhu, and Joseph Sackey, the cocoa farmer who allowed us to pitch our tents on his compound. In Boi-Tano, we are indebted to Samartex Timber Co. Ltd, especially to Messrs Andreas and Sarfo who accommodated us in their field camp. The help of the staff of the Forest Services Division, George Arthur and Frank Nyarko, are also duly acknowledged. In Krokosua Hills we were welcomed and supported by the people of Mmem village (located within the reserve), and Mr. Boamah of the Wildlife Division (Bia National Park).

Our local field assistants were of invaluable help during field work, as were the staff of the Forestry Commission, especially Mohammed Nuhu of Draw River, George Arthur and Frank Nyarko of Boi-Tano and Mr. Boamah of Krokosua Hills. We wish also to thank all the RAP participants for their pleasant and inspiring companionship. Mr. E.L Lamptey of the Wildlife Division issued the collecting and export permits. We are indebted to Conservation International (CI) in general and to Jennifer McCullough, Leslie Rice, Amy Heinemann and Leeanne E. Alonso in particular for the invitation to participate in and organization of this Rapid Assessment Program survey.

We wish to recognize the contributions of the Critical Ecosystem Partnership Fund as well as the Global Conservation Fund. Without their substantive and financial support, this project would not have been possible. Thanks is also due to members of CI's West Africa program, especially Jessica Donovan and Olivier Langrand, for input and guidance both before and after the expedition. We also owe thanks to Mark Denil of CI's Conservation Mapping Program, and to Kim Meek and Glenda Fabregas for their support in designing all RAP publications. We gratefully thank the Giuliani Family Foundation for providing funding for printing this report.

The invertebrate team is very grateful for the encouragement, the editorial advice given and the opportunity granted by Dr. D. K. Attuquayefio, Head of Department of Zoology, Legon for the opportunity to be part of the RAP expedition. The technical assistance of Mr. Ebenezer Adjedu also of Zoology Department Legon is acknowledged, as is the pleasant companionship of Mr. Patrick Ekpe, a colleague from Botany Department, University of Ghana, Legon, whose knowledge of forestry practices pulled the team from many sticky situations when official guides could not be found. Ms. Josephine T. Ahiati, Assistant Technician of Zoology Department, Legon, also deserves special thanks.

The small mammal team thanks farmer Kwadwo Collins in Mmem, Krokosua Hills, for information about cocoa farming. Thanks to Dr. Attuquayefio, University of Ghana, Zoology Department, for making available additional Sherman traps. Thanks to R. Hutterer at the Museum Alexander Koenig, Bonn, Germany, M. Carleton, L. Gordon and D. Smith at the United States National Museum, Washington, DC, and C. W. Kilpatrick at the University of Vermont for assistance with voucher specimen preservation and identification. We also thank Laurent Granjon, IRD, Bamako, Mali, for helpful comments on an earlier version of the manuscript.

The primate team wishes to express thanks to all of the guides and hunters who provided assistance during the surveys, especially Isaac Ochere, Peter Inkoon, Joseph Skido, Abraham Bindagim and Idrissu.

A biological assessment of the terrestrial ecosystems of the Draw River, Boi-Tano, Tano Nimiri and Krokosua Hills forest reserves, southwestern Ghana

5

# Participants and Authors

**Oscar Aalangdong** (large mammals)
Department of Renewable Natural Resources
University for Development Studies
P.O. Box 1350
Tamale, N.R.
GHANA
Email. irunuoh@yahoo.com

**Ebenezer Adjedu** (invertebrates)
Zoology Department
University of Ghana
P.O. Box LG67
Legon
GHANA
Email. ebenadjedu@yahoo.com

**Alex Agyei** (herps)
Wildlife Division
Forestry Commission
Box 47
Goaso, BA
GHANA
Email. alexaaaj@yahoo.com

**Frederick Ansah** (invertebrates)
Zoology Department
University of Ghana
P.O. Box LG67
Legon
GHANA
Email. frederickansah@hotmail.com

**Augustus Asamoah** (birds)
Ghana Wildlife Society
P.O. Box GP13252
Accra
GHANA
Email. aasamoah@mail.com

**Abdulai Barrie** (large mammals)
Department of Biological Sciences
Faculty of Environmental Sciences
Njala University College
University of Sierra Leone
PMB, Freetown
SIERRA LEONE
Email. ahbarrie@yahoo.com

**Dr. Jan Decher** (small mammals, scientific team leader)
Department of Biology
University of Vermont
Marsh Life Science
Burlington, VT 05405
UNITED STATES
Email. jan.decher@uvm.edu

**Tobias Deschner** (primates)
Max Planck Institute for Evolutionary Anthropology
Deutsche Platz 6
04163 Leipzig
GERMANY
Email. deschner@eva.mpg.de

**Patrick Ekpe** (plants)
Ghana Herbarium
Department of Botany
University of Ghana
P.O. Box LG 55
GHANA
Email. patrickekpe@yahoo.co.uk

**Raffael Ernst** (herps)
Department of Animal Ecology and Tropical Biology
Biocenter Am Hubland
97074 Würzburg
GERMANY
Email. ernst@biozentrum.uni-wuerzburg.de
       ernstraf@web.de

Present address:
Staatliches Museum für Naturkunde Stuttgart
Deptartment of Zoology
Rosenstein 1, 70191
Stuttgart
GERMANY

**Jakob Fahr** (contributing author)
Department of Experimental Ecology
University of Ulm
Albert-Einstein-Allee 11
89069 Ulm
GERMANY
Email. jakob.fahr@biologie.uni-ulm.de

**David Guba Kpelle** (primates)
Conservation International – Ghana
P.O. Box KAPT 30426
Accra
GHANA
Email. cioaa@conservation.org

**Torben Larsen** (butterflies)
UNDP VIETNAM
C/o Palais des Nations
1211 Geneva
SWITZERLAND
Email. torbenlarsen@compuserve.com

**Jennifer McCullough** (coordinator, editor)
Rapid Assessment Program
Conservation International
1919 M St. NW, Ste. 600
Washington, DC 20036
UNITED STATES
Email. j.mccullough@conservation.org

**James Oppong** (small mammals)
Wildlife Division
Forestry Commission
Box 47
Goaso, BA
GHANA
Email. james_oppongzooo@yahoo.com

**Hugo J. Rainey** (birds)
School of Biology
Bute Medical Buildings
University of St Andrews
St Andrews, Fife KY16 9TS
UNITED KINGDOM
Email. wcslactele@uuplus.com

**Mark-Oliver Rödel** (contributing author)
Department of Animal Ecology and Tropical Biology
Biocenter
Am Hubland, D-97074 Würzburg
GERMANY
Email. roedel@biozentrum.uni-wuerzburg.de

A biological assessment of the terrestrial ecosystems of the Draw River, Boi-Tano, Tano Nimiri and Krokosua Hills forest reserves, southwestern Ghana

7

# Organizational Profiles

## CONSERVATION INTERNATIONAL

Conservation International (CI) is an international, nonprofit organization based in Washington, DC. CI believes that the Earth's natural heritage must be maintained if future generations are to thrive spiritually, culturally, and economically. Our mission is to conserve the Earth's living heritage, our global biodiversity, and to demonstrate that human societies are able to live harmoniously with nature.

Conservation International
1919 M St., NW, Suite 600
Washington, DC 20036
USA
Tel.      1-202-912-1000
Fax.      1-202-912-0772
Web.      www.conservation.org
          www.biodiversityscience.org

## CONSERVATION INTERNATIONAL GHANA

Conservation International Ghana's work started in 1990 with the Kakum National Park, where the habitat of globally threatened species was secured against further degradation and species extinction through innovative ecotourism development. To further secure Kakum National Park, CI-Ghana implemented the Cocoa Agro-forestry Programme in partnership with Kuapa Kokoo, assisting cocoa farmers within the Kakum Conservation Area to adopt ecologically sustainable agronomic practices for increased production. This agroforestry initiative has provided a buffer zone and additional wildlife habitat for the threatened species within the park. As a result of CI-Ghana's interventions, Kakum National Park currently receives about 80,000 visitors annually, contributing significantly to the socio-economic development of Ghana.

From the project site at Kakum National Park, CI-Ghana has expanded its focus to the national level. CI-Ghana's work focuses on preventing species extinction, increasing protection and improving management of the remaining forest fragments, and the development of biodiversity corridors. To curb the threat of species extinction in Ghana, as a result of the bushmeat trade, CI-Ghana carried out a two-year nationwide bushmeat campaign. This was done in partnership with the Wildlife Division, Atomic Energy Commission, Ghana Standards Board, Food and Drugs Board, the Ministry of Food and Agriculture and the Environmental Protection Agency of Ghana. In partnership with the Ministry of Environment and Science, CI-Ghana provided technical support, secretariat and funding for the completion of the *National Biodiversity Strategy for Ghana*. To ensure the effective implementation of the strategy, CI-Ghana also provided technical support for the formulation of the action plan. Currently, CI-Ghana is represented on the National Biodiversity Committee in Ghana. In

December 1999, CI-Ghana facilitated a conservation priority-setting workshop that built a broad-based consensus on priorities for biodiversity conservation in the Upper Guinea forest ecosystem through active participation of 146 individuals from 90 institutions. Government, NGOs and private sector participants developed a common platform to guide and coordinate new investment and conservation at various scales throughout the region.

Conservation International Ghana
P.O. Box KAPT 30426
Airport, Accra
GHANA
Tel.     +233 21 / 780906
Fax.     +233 21 762009
Email.   oampadu-agyei@conservation.org

A biological assessment of the terrestrial ecosystems of the Draw River, Boi-Tano, Tano Nimiri and Krokosua Hills forest reserves, southwestern Ghana

9

# Report at a Glance

## A BIOLOGICAL ASSESSMENT OF THE TERRESTRIAL ECOSYSTEMS OF THE DRAW RIVER, BOI-TANO, TANO NIMIRI AND KROKOSUA HILLS FOREST RESERVES, SOUTHWESTERN GHANA

**Expedition Dates**
22 October-10 November 2003

**Area Description**
Four forest reserves make up the three sites surveyed during the RAP expedition: Draw River, Boi-Tano/Tano Nimiri (focusing primarily on Boi-Tano), and Krokosua Hills. Each of these four reserves contain areas designated as Globally Significant Biodiversity Areas (GSBAs) by the Government of Ghana. Draw River and Boi-Tano fall within the Wet Evergreen forest type and share the same characteristics. Krokosua Hills, which falls within the North-west subtype of Moist Semi-deciduous forest zone, has Upland Evergreen forest type occurring in steep-sided hills.

Draw River is situated in the southern portion of Ghana's high forest zone, and is contiguous to Nini-Suhien National Park and the Ankasa Resource Reserve (west of Draw River). Historically, logging has occurred in this forest reserve, and there are 40 ha of allowed farms. The forest consists of large trees with a vertically compressed canopy rarely exceeding 40 m. Old logging dumps and hauling tracks, relics of logging in 1990, are at various stages of recovery.

Draw River and Boi-Tano have similar patterns of vegetation and forest quality. The GSBA portion of Boi-Tano covers an eastern part of the forest in the southern half, including portions along the Tano River, and in the northern section of the forest. Non-GSBA compartments of this forest reserve have extensive networks of logging roads and are dominated by pioneer species.

The survey of Krokosua Hills was conducted from the small village of Mmem located within the reserve. The highest point within the reserve is 594 m. Krokosua Hills forest reserve consists of two forest blocks, one above and one below the Asempaneye-Kumasi road. The upper canopy consists of deciduous and evergreen species in varying proportions. Forest in most of the GSBA is slightly degraded, but some relatively good forest characteristic of this forest type remains.

**Reason for the Expedition**
The recent re-designation of 30 forest reserves as Globally Significant Biodiversity Areas (GSBAs) is a major innovation on the part of the Government of Ghana and presents an opportunity to conserve Ghana's biological resources of global conservation concern. In order to assist the Government of Ghana in assessing the faunal importance of these areas, to complement the established floral understanding, and to recommend appropriate conservation interventions to secure the ecological integrity of these key areas, Conservation International undertook a RAP survey of four areas designated as GSBAs.

**Major Results**
A variety of terrestrial habitats were observed during the RAP survey, including the Wet Evergreen forest type and the North-west subtype of Moist Semi-deciduous forest. Habitats were

evaluated with respect to plants (both woody and herbaceous), invertebrates, amphibians, reptiles, birds, and mammals (with special attention given to primates).

In total, the RAP team documented 1309 species with one amphibian species recorded being new to science. The team recorded range extensions for a number of species and added as new records for Ghana two plants, one butterfly, three amphibians and one shrew. The reserves investigated harbor a number of species of international conservation concern, including 24 "Black star" plant species; one reptile (*Osteolaemus tetraspis*) listed by IUCN as "Vulnerable"; five birds of global conservation concern including *Bleda eximia* and *Criniger olivaceus*, listed by IUCN as "Vulnerable"; two bat species listed by IUCN, one species as "Vulnerable" (*Scotophilus nucella*) and one species as "Near Threatened" (*Hipposideros fuliginosus*); one primate species listed by IUCN as "Endangered": *Pan troglodytes verus* (West African Chimpanzee) and one listed as "Near Threatened": *Procolobus verus* (Olive Colobus); and six additional large mammals of conservation concern including *Loxodonta africana cyclotis* (Forest elephant) listed as "Vulnerable" and *Cephalophus maxwelli, C. niger, C. dorsalis* and *Neotragus pygmaeus* listed as "Near Threatened".

## Number of Species Recorded:

| | |
|---|---|
| Plants | 681 |
| Invertebrates (non-butterfly) | 93 |
| Butterflies | 250 |
| Amphibians | 39 |
| Reptiles | 18 |
| Birds | 170 |
| Small mammals | 31 |
| Large mammals (excl. primates) | 20 |
| Primates | 6 |

## New species discovered:

| | |
|---|---|
| Amphibian | *Arthroleptis* sp. |

## New Records for Ghana:

| | |
|---|---|
| Plants | *Momordica sylvatica* |
| | *Tetrapleura chevalieri* |
| Butterfly | *Euriphene leonis* |
| Amphibian | *Acanthixalus* sp. |
| | *Arthroleptis* sp. |
| | *Phrynobatrachus annulatus* |
| Shrew | *Crocidura obscurior* |

## Species of Conservation Concern:

| | |
|---|---|
| Plants | 24 Black Star species |
| Reptiles | *Osteolaemus tetraspis* (VU) |
| | *Kinixys erosa* (DD) |
| Birds | *Bleda eximia* (VU) |
| | *Criniger olivaceus* (VU) |
| | *Bycanistes cylindricus* (NT) |
| | *Illadopsis rufescens* (NT) |
| | *Lamprotornis cupreocauda* (NT) |
| Bats | *Scotophilus nucella* (VU) |
| | *Hipposideros fuliginosus* (NT) |
| Large mammals | *Pan troglodytes verus* (VU) |
| | *Loxodonta africana cyclotis* (VU) |
| | *Cephalophus maxwelli* (NT) |
| | *Cephalophus niger* (NT) |
| | *Cephalophus dorsalis* (NT) |
| | *Neotragus pygmaeus* (NT) |
| | *Procolobus verus* (NT) |

## Conservation Recommendations

In order for the GSBAs to realize their potential for providing permanent 'stepping stones' for gene-flow, they must be connected to neighboring national parks and other reserves through corridors. Draw River should receive the highest level of protection and should be integrated as a part of Ankasa/Nini Suhien National Park, with which it is contiguous. The elephant population in this area migrates between Ankasa and Draw River and this population warrants further study, possibly through acoustic monitoring to determine population viability and distribution. Boi-Tano lies only a short distance from the northern edge of Nini Suhien NP. Consideration should be given to linking the two areas with a corridor of well-protected forest along natural corridors such as rivers. This should be wide enough to allow movement of a range of species.

In Krokosua Hills, the greatest priority for resource managers should be halting the spate of illegal clearance of forest for the cultivation of cocoa and embarking on innovative sustainable livelihood initiatives that are compatible with conservation of biodiversity. Conservation is likely to succeed when the people living close to biodiversity are aware of its value and able to enjoy the benefits from its preservation and sustainable utilization. Further surveys in areas where hunters claim to have spotted Miss Waldron's red colobus are strongly recommended. Confirmation of the presence of this monkey species could drastically speed up and facilitate the elevation of protection status and the acquisition of financial support through conservation agencies for these areas. Additionally, the presence of international researchers and local students is an effective way to secure protection of animals and reduce poaching activities.

# Executive Summary

As early as 1908, H.N. Thompson (1910) described alarming deforestation in Ghana. Shifting agriculture must have occurred for centuries, but the rate of deforestation accelerated about a century ago because of timber demands for newly mechanized gold mines, development of communications and a rapidly expanding area of farmland, including cocoa (Hawthorne and Abu Juam 1995). As far back as the 1920s and 1930s, foresters in Ghana demarcated and placed under management 280 forests for the purpose of ensuring the sustainable use of Ghana's forest resources and the preservation of forests with important roles as watersheds and windbreaks.

Ghana has lost roughly 80% of its forest habitat since the 1920s (Cleaver 1992) and about one-third of Ghana's forest is estimated to have disappeared in 17 years between 1955 and 1972 (Hall 1987). Of the original forest zone covering 82,260 km², the area under forest in 1973 amounted to 20,530 km² including 16,790 km² within forest reserves distributed more or less throughout the forest zone (Anon 1973). In 1988, forest cover in Ghana was estimated to be around 15,842 km² with the annual deforestation rate estimated at 220 km² (Sayer et al. 1992). Virtually all forests in reasonable condition today are in fact old designated reserves under the supervision of the Forest Services Division of the Forestry Commission. Many of these forests have retained a significant integrity, in the sense that the boundary lines laid down seventy years ago are still respected and the boundary lines regularly cleared and quite prominent. Various degrees of selective logging have taken place, but in most cases these logged areas maintain the possibility of regenerating to high forest. There are no real data but a rough estimate would be that about 10-12% of the original forest remains in at least moderate condition, with some pristine blocks in inaccessible areas.

## RAP EXPEDITION OVERVIEW AND OBJECTIVES

Together with CI-Ghana and CI's West Africa program, the Rapid Assessment Program (RAP) organized an expedition to four Globally Significant Biodiversity Areas (GSBAs) in southwest Ghana in October-November 2003 to better understand the biological diversity of this area. The primary objective of the RAP expedition was to collect scientific data on the diversity and status of species of these forest reserves in order to make recommendations regarding their protection and management. The forests investigated during the RAP expedition have been managed as productive forests, but were reclassified by the Government of Ghana as Globally Significant Biodiversity Areas (GSBAs). In 1999, the government of Ghana obtained the assistance of the Global Environment Facility (GEF) to implement legal establishment of GSBAs, designated based on Genetic Heat Index (GHI) of a reserve's botanical species. For the purpose of prioritizing plant conservation in Ghana, each plant species has been assigned to a star category, based on rarity. Black star species are internationally rare and uncommon in Ghana and urgent attention to the conservation of these species is called for. A high GHI signifies that an area is relatively rich in rare, black star species such that loss or degradation of the area would represent a highly significant erosion of genetic resources from the world, and Ghana in particular (Hawthorne and Abu-Juam 1995). In principle no logging or hunting should take place in GSBAs. However, it has never been the mandate of the Forest Services Division of the

Forestry Commission to implement hunting regulations and the Forestry Commission does not have the staff to police hunting within the GSBAs. Thus, larger mammals have become increasingly scarce. Whereas much information exists on Ghana's flora, surveys of fauna within these areas are incomplete, outdated, or non-existent. Our surveys focused on invertebrates, amphibians, reptiles, birds, and mammals to complement (and add to) the extensive botanical knowledge already collected for these sites, and to assess relative conservation importance.

The specific aims of the expedition were to:
- Derive a brief but thorough overview of the existing diversity and integrity within Draw River, Boi-Tano/Tano Nimiri, and Krokosua Hills GSBAs and evaluate their relative conservation importance;

- Undertake an evaluation of threats to the biodiversity within the areas surveyed;

- Provide on-site training under the guidance and mentorship of experienced field biologists and collaboration between Ghanaian and international scientists;

- Provide preliminary management recommendations for the surveyed areas together with recommendations for conservation priorities with a view to influencing local, national, and international conservation policies relating to southwest Ghana;

- Publish and make available to decision-makers and the general public the biodiversity and unique features that have been identified and described in the areas surveyed; and

- Increase awareness of the protected status of these GSBAs and promote their conservation.

The expedition's team of 14 scientists included representatives of three Ghanaian government organizations: the Wildlife Division of the Forestry Commission, the University of Ghana, and the University of Development Studies, as well as the Ghana Wildlife Society, a local NGO active in biodiversity conservation. The scientific team comprised international and national scientists specializing in West African terrestrial ecosystems and biodiversity.

## RESULTS BY SITE

The four forests surveyed belong to a group of 30 forest areas recently designated as Globally Significant Biodiversity Areas (GSBAs) and managed by Ghana's Forestry Commission (FC) under the Biodiversity component of the Natural Resource Management Program (NRMP) being funded by the GEF. The RAP survey took place from 22 October to 10 November 2003, at the end of the rainy season, and covered three sites:

**Draw River:** Wet evergreen forest type
The first RAP survey site (Site 1), Draw River, 5°12'N; 2°24'W, is situated in the southern portion of Ghana's high forest zone, and is contiguous to Ankasa National Park (which is west of Draw River). The zone of wet evergreen rainforest averages over 1,750 mm of rain per year (Hall and Swaine 1981). The reserve is divided into three blocks by rivers and the Ankrako-Kokum path. There are 40 ha of allowed farms within the reserve. Logging started in 1978 although no trees were removed between 1981 and 1985. In 1988 felling peaked at 712 trees, and figures after 1989 (332 trees) are not available. There is much variation in genetic heat in this reserve depending on soil and landform, though even the "coolest" areas have a high genetic heat (i.e. rich in rare plant species).

The forest consists of large trees with a vertically compressed canopy rarely exceeding 40 m. Old logging dumps and hauling tracks, relics of logging in 1990, are at various stages of recovery. Human interference through small scale harvesting of minor forest products (canes, raphia palm, bamboo, trapping for game) is noticeable in many areas. Draw River ranks 8th of all forests in Ghana with regards to Genetic Heat Index (GHI).

Significant findings:
- We recorded nine black star species.

- The butterfly fauna of Draw River contains many endemics to the Ghana subregion and to West Africa, including a large number of rare species. We observed a number of indicator species of wetter forest systems in good condition. The rare *Pteroteinon pruna* was found, probably an indication that a number of rare species known from Ankasa are also present in Draw River.

- For amphibians, we recorded *Geotrypetes seraphini*, a rarely recorded Caecilian of uncertain conservation status, as well as a recently described Upper Guinean endemic ranid frog, *Phrynobatrachus phyllophilus* and a new species of the genus Arthroleptis. There is a high resemblance in amphibian species composition between Draw River and Taï National Park (Côte d'Ivoire). The vast majority of species recorded were either Upper Guinean rain forest endemics or at least restricted to West Africa.

- We recorded two reptile species listed under CITES Appendices I and II: *Osteolaemus tetraspis* (I) and *Chamaeleo gracilis* (II).

- We recorded 126 bird species including 3 of conservation concern: *Bleda eximia* (Vulnerable), *Criniger olivaceus* (Vulnerable), and *Lamprotornis cupreocauda* (Near Threatened). According to our results, Draw River holds more forest biome species than Ankasa and Nini-Suhien combined. Relatively few forest edge or farmbush species were recorded. Few large species, such as hornbills, were recorded possibly indicating that large birds have been targeted by hunters.

A biological assessment of the terrestrial ecosystems of the Draw River, Boi-Tano, Tano Nimiri and Krokosua Hills forest reserves, southwestern Ghana

13

- We recorded 19 species of large mammals including *Dendrohyrax dorsalis* Western tree hyrax, *Cephalophus niger* Black duiker (Near Threatened), *Cephalophus dorsalis* Bay duiker (Near Threatened), *Anomalurus beecrofti* Beecroft's anomalure, *Neotragus pygmaeus* Royal antelope (Near Threatened), and *Hylochoerus meintzhageni* Giant forest hog, as well as the *Loxodonta africana cyclotis* Forest elephant (Vulnerable).

- We confirmed the presence of *Galagoides demidoff* Demidoff's Galago, *Perodicticus potto* Potto, *Cercopithecus m. lowei* Lowe's guenon, *Cercopithecus c. petaurista* the Lesser spot-nosed guenon, and *Procolobus verus* Olive Colobus (Near Threatened) and we found feeding remains of *Pan troglodytes verus* West African chimpanzee (Endangered).

- The shrews *Crocidura buettikoferi* and *C. foxi* were recorded for the first time from this part of Ghana.

**Boi-Tano (/Tano Nimiri):** Wet evergreen forest type
The second RAP survey site (Site 2) was Boi-Tano/Tano Nimiri forest reserves, focusing mostly on Boi-Tano 5°32'N; 2°37'W. Access in many areas here was difficult due to watercourses and swamps. The GSBA covers an eastern part of the forest in the southern half, including portions along the Tano River, and in the northern section of the forest. Logging of this forest progressed rapidly in the late 1990s, and by 2002 all non-GSBA compartments from Samreboi to the southwestern tip were opened up, forming an extensive logging road network clearly visible on satellite imagery (Hawthorne 2002). The forest here has sample GHIs often in excess of 250. The Tano River and larger tributaries support communities of riverine species with considerable variation between ridge-top and low-lying forests.

The forest here has similar characteristics to Draw River, with a pattern of vegetation and forest quality suggesting that this forest is also still in healthy condition. Non-GSBA compartments are dotted with logging scars and dominated by pioneer species. Some areas outside of the GSBA are severely affected by log dumps and road verges. The main hauling road appears unusually wide (wider than allowed by forestry regulations) and is lined with *Cercropia peltata*, an introduced pioneer species. Boi-Tano ranks 5th with regards to GHI for all Ghanaian forests.

Significant Findings:
- Formerly 21 black star plant species were known from the Boi-Tano GSBA, 11 of which were recorded during our survey. We also added two additional black star species known from this site during our survey: *Momordica sylvatica* and *Tetrapleura chevalieri*. These two records are new for Ghana.

- With regards to butterflies, a number of positive indicator species for wetter rainforests in good conditions were present, despite poor collecting conditions and limited material recorded. The rare Hesperiid, *Hypoleucis sophia*, was present – there are probably fewer than 20 in collections anywhere.

- There is a high resemblance in amphibian species composition between Boi-Tano and Taï National Park. The vast majority of species recorded were either Upper Guinean rain forest endemics or at least restricted to West Africa. The relatively high diversity and/or unique species composition with respect to regional endemicity clearly demonstrates that these areas still have a high potential for conservation.

- We recorded 109 bird species in Boi-Tano. Of these, two were of conservation concern: *Bleda eximia* (Vulnerable) and *Illadopsis rufescens* (Near Threatened). Boi-Tano now ranks second only to Kakum National Park in Ghana for number of forest biome species recorded. Relatively few forest edge or farmbush species were recorded. Few large species, such as hornbills, were recorded.

- We recorded 14 species of large mammals in Boi-Tano/Tano Nimiri. *Dendrohyrax dorsalis* Western tree hyrax, *Cephalophus niger* Black duiker (Near Threatened), Bay duiker *Cephalophus dorsalis* (Near Threatened), and *Neotragus pygmaeus* Royal antelope (Near Threatened) were recorded.

- At Boi-Tano and Tano Nimiri no diurnal monkey presence could be confirmed through visual or acoustical signs. We confirmed the presence of two prosimian species (*Galagoides demidoff* Demidoff's Galago and the *Perodicticus potto* Potto) and found one nest group consisting of five chimpanzee nests at Tano-Nimiri. Additionally we found feeding remains of *Chrysophyllum subnudum* fruits with chimpanzee teeth imprints.

**Krokosua Hills:** North-west subtype of Moist Semideciduous forest
The third RAP survey site (Site 3) was Krokosua Hills Forest Reserve 6°37'N; 2°51'W. Our surveys were conducted from the small village of Mmem (population around 75) legally remaining several kilometers within the reserve boundary. The highest point within the reserve is 594 m. Krokosua Hills Forest Reserve consists of two forest blocks, one above and one below the Asempaneye-Kumasi road. The southern portion of the reserve is heavily hunted and structurally degraded and contains numerous agricultural plots (Magnuson 2002). During the RAP survey, only the northwestern portion of the reserve (corresponding to the GSBA) was surveyed. Because this portion has a relatively small human population and more dramatic topography, the forest here has remained in a relatively more intact state than the southern portion (Magnuson 2002). Krokosua receives 1,250-1,500 mm rain per year. This range of hills includes high, flat-topped bauxite capped hills of Nsesreso, claimed by

locals to be an abandoned mining village. Within the reserve there are a number of farms, old mine pits, and villages. Alluvial gold mining was carried out in the southwestern portion in the past (Hawthorne 2002). A botanical survey in 1991 confirmed extreme patchiness with very damaged, logged areas on flatter land and unlogged areas with less damage on steep slopes. The GSBA was last logged in 1988. Chain-saw activity, mainly in the night, poses a major problem in the area (and was observed during the RAP).

Trees in this type of forest become taller than those of other Ghanaian forest types; heights often exceed 50 to 60 m. Upper canopy consists of deciduous and evergreen species in varying proportions. Forest in most of the GSBA is slightly degraded but with relatively good forest characteristic of this forest type. Portions east of Mmem are degraded and are recovering from bush fires, which swept the forest in 1993. Relics of chain-sawed lumber are common.

This GSBA is currently threatened by increasing encroachment through illegal clearing of the best part of the forest for farming purposes by inhabitants of Mmem. Nine pockets of illegal farms, ranging from 0.5 to 1.5 ha were encountered during our survey. Conversations with villagers revealed that people from fringing communities are moving into Mmem where they are allocated forestland for farming under shared cropping agreements with the village chief. Threats include considerable encroachment of cocoa and high hunting pressure. There is still illegal logging (especially with chainsaws) and legal logging has been undertaken in a less sensitive fashion than in the other GSBAs surveyed. The central ridge still contains intact forest, and is in far better condition than lower adjacent forests. Much of Krokosua Hills is heavily logged and access roads are wide and clogged with the invasive weed *Chromolaena odorata*. This presents additional problems during the dry season as the dried weed increases the fire hazard to the forest.

Significant Findings:

- Along the ridge we found *Eruiphene leonis* for the first time in Ghana – this butterfly species is known, but rare, in Liberia and Sierra Leone, and also very scarcely known in Côte d'Ivoire. Also recorded was *Catuna niji*, previously known in Ghana only from Ankasa.

- Amphibian communities of Krokosua Hills have strong species affiliations with Mt. Peko National Park in Côte d'Ivoire. The vast majority of species recorded were either Upper Guinean rain forest endemics or at least restricted to West Africa. We recorded the first country record for the genus *Acanthixalus*, as well as a new species of the family Arthroleptidae.

- We recorded 138 bird species in Krokosua Hills. Of these, four were of conservation concern: *Bycanistes cylindricus* (Near Threatened), *Bleda eximia* (Vul-

nerable), *Illadopsis rufescens* (Near Threatened) and *Lamprotornis cupreocauda* (Near Threatened). Krokosua now qualifies as an Important Bird Area (as designated by BirdLife International). Relatively few forest edge or farmbush species were recorded. Few large species were recorded.

- We recorded 14 species of large mammals in Krokosua including *Dendrohyrax dorsalis* Western tree hyrax, *Cephalophus niger* Black duiker (Near Threatened), *Cephalophus dorsalis* Bay duiker (Near Threatened), *Neotragus pygmaeus* Royal antelope (Near Threatened), and *Anomalurus beecrofti*, Beecroft's anomalure.

- At Krokosua Hills we confirmed the presence of two prosimian species (*Galagoides demidoff* Demidoff's galago and *Perodicticus potto* Potto) as well as *Cercopithecus m. lowei* Lowe's guenon, and *Cercopithecus c. petaurista* the Lesser spot-nosed guenon.

- At Krokosua Hills we found relatively high bat diversity and abundance including a population of the bat *Scotophilus nucella*, which was described in 1984 and known from only ten other specimens.

## RESULTS BY TAXON

**Plants.** We recorded 681 vascular plant species belonging to 94 families. Of 20 Black star species known from Draw River, we recorded nine during our survey. Formerly 21 black star species were known from the Boi-Tano GSBA, 11 of which were recorded during our survey. We also added two additional Black star species known from this site during our survey: *Momordica sylvatica* and *Tetrapleura chevalieri*. These two records are also new for Ghana. Three black star species and a number of gold star species are known from Krokosua Hills. We recorded all three, in addition to a gold star species, *Hymenostegia aubrevellei*, known only from this GSBA within Ghana.

**Butterflies.** The butterfly fauna of Draw River contains many endemics to the Ghana subregion and to West Africa, including a large number of rare species. Draw River produced a number of indicator species of the wetter forest systems in good condition, including *Euriphene veronica*, previously recorded from Ghana only in Ankasa. A single female of the rare *Pteroteinon pruna* was found, probably an indicator that a number of rare species known from Ankasa are also present in Draw River. At Boi-Tano, a number of positive indicator species for wetter rainforests in good conditions were present, despite poor collecting conditions and limited material recorded. The rare hesperiid, *Hypoleucis Sophia*, was present – there are probably fewer than 20 in collections anywhere. Krokosua Hills, while degraded in some places, still contains a number of areas with good forest. Along the ridge we found *Eruiphene leonis* for the first time in Ghana – this species is known, but rare, in Liberia and Sierra Leone, and also very scarcely known in Côte d'Ivoire. Also recorded was *Catuna niji*, previously known in Ghana only from Ankasa. The general impression is that Draw River and Boi-Tano are effective reservoirs of the West African wet forest fauna. The hilly parts of Krokosua have a varied and interesting fauna.

**Amphibians.** We recorded 39 amphibian species, among them the first country record for the genus *Acanthixalus* (known only from Côte d'Ivoire and Cameroon), as well as a new species of the genus *Arthroleptis*. In Draw River we recorded *Geotrypetes seraphini*, a rarely recorded Caecilian of uncertain conservation status, as well as the endemic Ghanaian ranid, *Phrynobatrachus ghanensis,* which was previously only known from Kakum NP and Bobiri Forest Reserve. We recorded *Phrynobatrachus annulatus* for the first time in Ghana (from both Draw River and Boi-Tano). An *Astylosternus* tadpole from Draw River may represent *Astylosternus occidentalis* or an undescribed species. The species richness of the forests under investigation can be considered moderate to high. This is especially true when compared to the most diverse West African region so far investigated, Taï National Park. Although the recorded species richness in Draw River and Boi-Tano does not reach the same level as Taï, there is a high resemblance in species composition. Communities of Krokosua Hills have strong species affiliations with Mt. Peko National Park. In all three forests the vast majority of species recorded were either Upper Guinean rain forest endemics or at least restricted to West Africa. The relatively high diversity and/or unique species composition with respect to regional endemicity clearly demonstrates that these areas still have a high potential for conservation. However, we also documented several farmbush or even savannah species, normally not occurring in forest areas, which is a clear indication of alteration of original forest habitat through unsustainable forest use (encroachment, illegal farming, and timber extraction).

**Reptiles.** We recorded 18 reptile species including three of global conservation concern (*Chamaeleo gracilis, Osteolaemus tetraspis,* and *Kinixys erosa*). We believe these results underrepresent actual numbers due to low detectibility in the field – especially for snakes. Nevertheless, four of the six snake species recorded are endemic to the Upper Guinea forest. These include the poorly known *Lycophidion nigromaculatum, Dendroaspis viridis,* and *Thrasops* sp. Sightings of chelonians and crocodilians were rare and local villagers reported hunting the African dwarf crocodile, *Osteolaemus tetraspis,* and Forest hinged tortoise, *Kinixys erosa,* for food.

**Birds.** We recorded 170 bird species, 126 in Draw River, 109 in Boi-Tano, and 138 in Krokosua Hills. Of these, five were of conservation concern (three in Draw River: *Bleda eximia, Criniger olivaceus,* and *Lamprotornis cupreocauda*; two in Boi-Tano: *Bleda eximia* and *Illadopsis rufescens*; and four in Krokosua Hills: *Bycanistes cylindricus, Bleda eximia, Illadopsis rufescens* and *Lamprotornis cupreocauda*). A substantial component of the forest-restricted species in the country was found, thus placing all three sites in the top six forests in Ghana for numbers of forest birds. Krokosua now qualifies as an Important Bird Area (as designated by BirdLife International), joining Draw River and Boi-Tano, which our surveys have found to be more important than previously thought. According to our results, Draw River holds more forest biome species than Ankasa and Nini-Suhien combined. Boi-Tano now ranks second only to Kakum National Park in number of forest biome species recorded. Relatively few forest edge or farmbush species were recorded at any site, further indicating the quality of the forests. Few large species were recorded at any site. Guineafowl, large birds of prey, *Corythaeola cristata* Great Blue Turacos and large hornbills were rarely recorded if at all. It is possible that these large birds have been targeted by hunters and that their populations are now much reduced.

**Small mammals.** One hundred and six (106) terrestrial small mammal captures, composed of six species of shrews and eight species of rodents, and 82 bat captures composed of 15 species were made. The shrew *Crocidura obscurior*, caught at all three sites, was the first record of this species for Ghana. *Crocidura buettikoferi* and *C. foxi* are first records for this region of Ghana. At Krokosua Hills we found a population of the rare microbat *Scotophilus nucella*, which was described in 1984 and hitherto known from only ten specimens from Ghana, Côte d'Ivoire and Uganda. Overall the small mammal species composition clearly reflects a forest fauna. Especially among the bats sampled, not a single savannah species was present, despite partially degraded forest conditions.

**Large mammals.** We recorded 19, 14 and 14 species of large mammals in Draw River, Boi-Tano/Tano Nimiri, and Krokosua Hills respectively. *Dendrohyrax dorsalis* Western tree hyrax, *Cephalophus niger* Black duiker (Near Threatened), *Cephalophus dorsalis* Bay duiker (Near Threatened), and *Neotragus pygmaeus* Royal antelope (Near Threatened) were recorded from all sites. Using, track, dung and other signs, we documented the presence of *Loxodonta africana cyclotis,* the Vulnerable Forest elephant in the Draw River forest reserve. *Anomalurus beecrofti* Beecroft's anomalure was recorded at Draw River and Krokosua Hills. No Yellow-backed duikers, Leopards, Aardvark, Pangolins or Bongo were observed but local hunters reported that these species still occur in these forest reserves.

**Primates.** Overall, we confirmed the presence of two prosimian species (*Galagoides demidoff* Demidoff's galago and *Perodicticus potto* Potto) for all sites and three anthropoid monkey species (*Cercopithecus campbelli lowei* Lowe's monkey, *Cercopithecus petaurista* Lesser spot-nosed guenon and *Procolobus verus* Olive colobus) for one or more sites. The presence of *Pan troglodytes verus* West African Chimpanzee could be confirmed for two sites (Draw River and Tano

Nimiri). Interviews with local hunters indicate the presence of additional primate species (*Cercopithecus diana roloway* Roloway's guenon, *Colobus polykomos vellerosus* White-thighed black and white colobus and *Cercocebus atys lunulatus* White-naped mangabey. At three of the sites hunters claimed to have recently seen individuals of *Procolobus badius waldronii* Miss Waldron's red colobus, a species considered to be extinct (Oates et al. 2000).

At Draw River, we visually confirmed the presence of Lowe's guenon, the Lesser spot-nosed guenon, and the Olive colobus. At Boi-Tano and Tano Nimiri no diurnal monkey presence could be confirmed through visual or acoustical signs. At Krokosua Hills, we visually confirmed the presence of Lowe's guenon and the Lesser spot-nosed guenon. We found feeding remains of chimpanzees (piles of *Sacoglottis gabonensis* kernels with food wedges) at Draw River. We found one nest group consisting of five chimpanzee nests at Tano-Nimiri. Additionally we found feeding remains of *Chrysophyllum subnudum* fruits with teeth imprints of chimpanzee. At Krokosua Hills we found chimpanzee feeding remains of *Marantochlea leucantha* stems.

## CONSERVATION RECOMMENDATIONS

It is worth considering the size of the present national parks: Ankasa/Nini Suhien is 509 km², Bia National Park/Bia Resource Reserve cover an area of 306 km², and Kakum is 370 km². While forests such as Nini-Suhien and Kakum may be large enough to maintain genetically viable populations of butterflies and most arthropods or even many small mammal species, this might not be so for the larger inhabitants, such as forest buffalo or elephant. All of the national parks at present have a potential gene-flow with some neighbouring forests in reasonable condition. The GSBAs have the potential for providing permanent 'stepping stones' for gene-flow.

Priority Actions for Conservation:

- Draw River should receive the highest level of protection and should be integrated as a part of Ankasa/Nini Suhien NP, with which it is contiguous. The elephant population in this area migrates between Ankasa and Draw River and this population warrants further study, possibly through acoustic monitoring to determine population viability and distribution.

- Boi-Tano lies only a short distance from the northern edge of Nini Suhien NP. Consideration should be given to linking the two areas with a corridor of well-protected forest along natural corridors such as rivers. This should be wide enough to allow movement of a range of species. This could be important for a number of species that make long distance movements such as hornbills, parrots and elephants. Boi-Tano should receive increased protected area status if linked to Nini Suhien NP as it is one of the few known sites for a number of

threatened species including *Agelastes meleagrides* White-breasted Guineafowl.

- Reconsider zones for logging within the Boi-Tano and Tano Nimiri GSBAs, especially logging that would lead to further habitat fragmentation and destruction of confirmed chimpanzee habitat. These two GSBAs are within adjacent forest reserves and should be linked by sufficient corridors of good quality forest.

- The Forestry Commission should hold a stakeholders' forum to decide the status of Krokosua Hills and to clarify who is in charge of day-to-day management. FSD personnel appear to have been replaced by WD staff. Perhaps related to the fact that while most GSBAs were declared based on genetic heat index (GHI), this GSBA was selected due to the fact that this site was a former refuge for Miss Waldron's red colobus and it is hoped that the species may still be found here.

- At sites where hunters repeatedly claimed recent sightings of Miss Waldron's red colobus (Boi-Tano, Tano Nimiri and Krokosua Hills) further surveys are strongly recommended. Confirmation of the presence of this monkey species could drastically speed up and facilitate the elevation of protection status and the acquisition of financial support through conservation agencies for these areas.

General Conservation Recommendations:

- Monitor and enforce penalties for the use of illegal snares that are found by the dozens along trails, resulting in indiscriminate killing of wildlife.

- Strictly enforce logging regulations (including the clearing of forest for cocoa plantations), especially in the areas designated Globally Significant Biodiversity Areas (GSBAs).

- Villages permitted in the forest for historical or political reasons, such as the village of Mmem (Mim) at Krokosua need to strictly adhere to the original forest boundaries determined around them. Villages with such exceptional status inside protected forests should receive special attention to develop alternative economic opportunities, such as sustainable harvesting and processing of non-timber forest products (NTFPs), gainful involvement in local law enforcement, and the potentially profitable development of facilities for ecotourism (guest houses, visitor's center, local guides etc.).

- Educational posters or other learning materials need to be made available in schools and communities to promote an understanding of conservation and biodiversity. Chimpanzees, due to their acknowledged closeness to humans, could play a vital role in educational campaigns.

- Establish wildlife guard camps in all areas and train guards in patrolling techniques. A regular monitoring of guard activities is necessary, as well as continuous motivation and support.

- Attract international researchers and local students to establish long-term research sites in the areas. A growing body of experience clearly indicates that permanent presence of researchers is an effective way to secure protection of animals and reduce poaching activities.

- Connect GSBAs to neighboring national parks and other reserves through corridors.

- More funds and an increased focus on wildlife monitoring and protection are necessary. Implement a total ban on hunting in some forests in Ghana to create refuge and recovery areas for certain species, from whence other areas can be re-populated.

- FSD staff attempted to re-plant two illegal farms in Krokosua using *Cedrela odorata*, a non-native species. Any replanting scheme in a GSBA should use species native to that particular forest.

- In the long run, a relocation of the Mmem village which is within the forest reserve has to be seriously considered, since immigration into this village is high and ongoing, leading to an increase in demand for land and consequently to a high conflict potential between villagers and wildlife guards. Alternatively, Mmem could become a conservation outpost with a research station, ecotourism guest house, and only restricted and sustainable cocoa farming in a clearly defined buffer zone, using shade tolerant varieties, and NTFP harvesting and processing (basketry, medicinal plants etc.), perhaps as a model of how a village could live in harmony inside a reserve and assist the reserve.

- Survey of the internal boundary of Mmem village needs to be conducted as soon as possible.

- In Boi-Tano, the central access road, an old logging road, is excessively broad now that there is no heavy traffic except seasonal cocoa trucks. The roads impede many arthropods from crossing from one part of the forest to the other and allow the growth of the invasive weedy plant *Chromolaena odorata*; closing the canopy above the road would eliminate this noxious weed.

## REFERENCES

Anon. 1973. Annual Report of the Forestry Department, Ministry of Lands and Mineral Resources, Ghana, for 1972. Forestry Division, Accra.

Cleaver, K. 1992. Deforestation in the western and central African rainforest: the agricultural and demographic causes, and some solutions. Pages 65-78. *In:* Cleaver, K., M. Munasinghe, M. Dyson, N. Egli, A. Penker and F. Wencelius (eds.). Conservation of West and Central African Rainforests. The World Bank/International Union for the Conservation of Nature, Washington, DC. 351 pp.

Hall, J.B. 1987. Conservation of Forests in Ghana. Universitas. 8:33-42. University of Ghana, Legon, Ghana.

Hall, J.B. and M.D. Swaine. 1981. Distribution and ecology of vascular plants in a tropical rain forest. Forest vegetation in Ghana. Geobotany 1. Junk, The Hague. 383 pp.

Hawthorne, W.D. 2002. Final report of the floral survey of the Biodiversity Component of NRMP. Forestry Commission, Biodiversity Conservation Component. Ministry of Lands and Forestry, Ghana.

Hawthorne, W.D. and M. Abu-Juam. 1995. Forest Protection in Ghana. IUCN/ODA/Forest Department Republic of Ghana, Gland, Switzerland, and Cambridge, UK. xvii + 203 pp.

Magnuson, L.E. 2002. Distribution and Habitat Use of the Roloway Guenon (*Cercopithecus Diana roloway*) in Ghana, West Africa. Master's thesis, Natural Resources: Wildlife Management, Humboldt State University. 68 pp.

Oates, J.F., M. Abedi-Lartey, S. McGraw, T.T. Struhsacker and G.H. Whitesides. 2000. Extinction of a West African red colobus monkey. Conservation Biology, 14:1526-1532.

Sayer, J.A., C.S. Harcourt and N.M. Collins. 1992. The conservation atlas of tropical forests. Africa. IUCN, Macmillan, New York, 335pp.

Thompson, H.N. 1910. Gold Coast: report on forests. Colon. Rep. Miscell. 66:1-238.

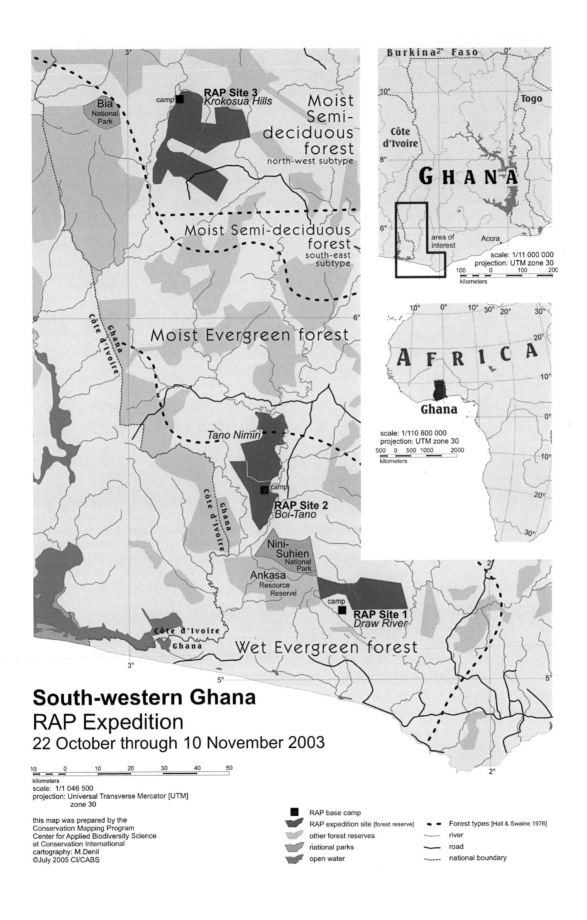

**RAP Site 3**
*Krokosua Hills*

Moist Semi-deciduous forest
north-west subtype

Moist Semi-deciduous forest
south-east subtype

Moist Evergreen forest

*Tano Nimiri*

**RAP Site 2**
*Boi-Tano*

Nini-Suhien
National Park

Ankasa
Resource
Reserve

**RAP Site 1**
*Draw River*

Wet Evergreen forest

Bia
National
Park

Burkina Faso
Togo
Côte d'Ivoire
GHANA
area of interest
Accra

scale: 1/11 000 000
projection: UTM zone 30
100    0    100    200
kilometers

AFRICA
Ghana

scale: 1/110 800 000
projection: UTM zone 30
500  0  500 1000    2000
kilometers

# South-western Ghana
## RAP Expedition
### 22 October through 10 November 2003

10    0    10    20    30    40    50
kilometers
scale: 1/1 046 500
projection: Universal Transverse Mercator [UTM]
zone 30

this map was prepared by the
Conservation Mapping Program
Center for Applied Biodiversity Science
at Conservation International
cartography: M.Denil
©July 2005 CI/CABS

■ RAP base camp
RAP expedition site [forest reserve]
other forest reserves
national parks
open water

Forest types [Hall & Swaine 1976]
river
road
national boundary

(Photo: J. Decher)

(Photo: T. Larsen)

Little collared fruit bat (*Myonycteris torquata*) from Draw River Forest Reserve.

Local hunter with Maxwell's Duiker (*Cephalophus maxwelli*) at Krokosua Hills Forest Reserve.

(Photo: J. Decher)

Local forestry officers with Roloway monkey (*Cercopithecus diana roloway*) skin at Mmem Village, Krokosua Hills Forest Reserve.

(Photo: J. Decher)

Forest edge adjacent to recently clearcut and burned farm plot near Mmem Village inside Krokosua Hills Forest Reserve.

(Photo: J. Decher)

Nagtglas's Dormouse (*Graphiurus nagtglasii*) caught at base of hollow tree at Krokosua Hills Forest Reserve.

(Photo: R. Ernst)

The frog, *Phrynobatrachus calcaratus*, a widespread and locally abundant species.

(Photo: J. McCullough)

Access road at Krokosua Hills Forest Reserve.

(Photo: J. McCullough)

Entomologist Torben Larsen checking butterfly trap at Krokosua Hills Forest Reserve.

A biological assessment of the terrestrial ecosystems of the Draw River, Boi-Tano, Tano Nimiri and Krokosua Hills forest reserves, southwestern Ghana

21

(Photo: J. Decher)

International Ghana RAP team at Mmem Village, Krokosua Hills Forest Reserve.

(Photo: J. Decher)

David Guba Kpelle (CI Ghana) and primatologist Tobias Deschner conducting local knowledge interviews using pictures of primate species at Mmem Village, Krokosua Hills.

(Photo: J. Decher)

Mouse-tailed Shrew (*Crocidura muricauda*) from Boi-Tano Forest Reserve.

# Chapter 1

## An ecological, socio-economic and conservation overview of Draw River, Boi-Tano, Tano Nimiri and Krokosua Hills forest reserves, southwestern Ghana

*Jan Decher, David Guba Kpelle and Jennifer McCullough*

## INTRODUCTION

Current biodiversity patterns and plant and animal endemism in the Upper Guinea Forest most likely are the result of repeated climate changes during the Pleistocene epoch, 10,000-1.9 million years B.P. The ensuing dry conditions in the tropics created isolated forest refugia, some of which have been identified as centers of endemism (Grubb 1982). Repeated expansions and contractions of the original forest led to genetic differentiation and speciation of the fauna and flora. At its greatest expansion since the last glaciation peak (18,000 years before present) the Upper Guinea Forest ecosystem is estimated to have covered as much as 700,000 km² (Myers 1982, Hamilton and Taylor 1991), but centuries of human activity have resulted in the loss of nearly 70% of the original forest cover (Bakarr et al. 2001). The remaining Upper Guinea Forest is restricted to a number of more or less disconnected forest reserves and a few national parks acting as man-made refuges for the region's unique species of flora and fauna. These remaining forests still contain exceptionally diverse ecological communities, distinctive flora and fauna, and several forest types harboring a number of endemic species.

Site-specific studies conducted in different countries within the Upper Guinea Forest show high levels of local endemism in plant species, which is likely reflective of the region as a whole (Bakarr et al. 2001). Of 1,300 species known from Taï National Park in Côte d'Ivoire, nearly 700 are confined to the Upper Guinea Forest ecosystem and 150 are endemic to Taï (Davis et al. 1994). On Mount Nimba, bordering Guinea, Côte d'Ivoire and Liberia, 2,000 plant species are estimated to occur in an area of only 480 km², with 13 being endemic to that site.

While the Upper Guinea Forest is home to a unique assemblage of ecosystems, it also houses vast stores of valuable natural resources, and all within a region facing extreme poverty, growing human population densities, weak environmental governance, habitat loss and in some parts periodic, but persistent, civil unrest. Determining how to best make use of the biological and geological resources, to the advantage of local communities and regional biodiversity, is a complex challenge faced by West African governments, development organizations, conservationists, NGOs, extractive industries and local inhabitants.

Southwestern Ghana, specifically the section including Draw River, Boi-Tano and Tano Nimiri forest reserves, has the highest botanical diversity within Ghana as measured by the Genetic Heat Index (GHI) used by the FD/ODA Forest Inventory and Management Project (FIMP) in Kumasi (Hawthorne 1996). For butterflies close to 400 species are to be expected for this area, whilst for mammals, southwestern Ghana supports the majority of the 229 plus species known from Ghana (Grubb et al. 1998).

### Geology

West Africa, and thus Ghana, geologically belongs to the very old (570 to 4,600 million years) Precambrian Guinean shield of the former supercontinent Gondwana. Southwestern Ghana is dominated by Precambrian rock formations of the Birimian (Middle Pre-Cambrian) and Tarkwaian (Upper Pre-Cambrian) series which are of special economic significance due to their deposits of gold, diamonds and manganese. The southwestern part of Ghana was eroded into a peneplain, subsequently dissected into hills and valleys or flat uplands separated by valleys (Furon 1963, Gnielinski 1986, Grubb et al. 1998).

## Geography

Ghana lies along the Gulf of Guinea in West Africa and is located within longitudes $3^0$ 5'W and $1^0$10'E and latitudes $4^0$ 35' N and $11^0$N. It covers an area of about 239,000 km², and is bordered by Burkina Faso, Côte d'Ivoire and Togo. Ghana can be divided into five natural regions: the coastal or *low plains*, as a broad belt along the Gulf of Guinea, the *Ashanti highlands* to the northwest, the *Akwapim-Togo mountains* in the East, the *Volta basin* and the plateans and terraces of the *high plains* in the north of the country (Gnielinski 1986). The southwestern low plains or Akan lowlands, which include the Draw River, Boi-Tano, and Tano Nimiri forest reserves, consist of the river basins of the Tano and Ankobra Rivers in the west and the Pra and Densu rivers in the east. Several hill chains running from the northeast to the southwest define the river basins, rarely exceeding 300 m elevation with a few peaks reaching 500 m. The Tano and Ankobra basin forms the largest area of the Akan lowlands including the gold mines of Tarkwa, the diamond fields of the Bonsa valley and the manganese deposits near Nsuta. Significant bauxite reserves are also present (Gnielinski 1986). Additionally, Ghana can be divided into five biogeographical zones: the Guineo-Congolian, including the wet evergreen and moist semi-deciduous, in the southwest, the Guineo-Congolian-Sudanian transitional zone in the middle and the south-east, the Sudanian in the north and the Sub-Sahelian in the northeastern corner (Ministry of Environment and Science 2002).

## Climate

Compared to many major rainforest areas around the world, the Guinean Rainforest is drier and seasonality is more pronounced (Davis et al. 1994). Most of the Guineo-Congolian Region receives between 1,600 and 2,000 mm of rainfall per year, which is less evenly distributed than that of other rainforests (White 1983). In the main eastern block of the Guineo-Congolian Region, in general, rainfall shows two peaks separated by one relatively severe and one less severe dry period (White 1983).

Throughout the Guineo-Congolian Region, average monthly temperature is almost constant throughout the year with a mean annual temperature in Ghana of 26°-29° C and a daily range along the coast of only 6° to 8° C (White 1983, Grubb et al. 1998).

In this area, two rainy seasons occur from about April to June and from September to November separated by a "little dry season" of about six weeks during July and August. This pattern corresponds to the two seasonal latitudinal passes of an "equatorial trough of low pressure encircling the earth near the equator" (Martin 1991:49) more formally known as the Intertropical Convergence Zone (ITC) over the African landmass (Ojo 1977). Annual rainfall ranges from about 750 mm in the northern forests to over 1,750 mm in the southwestern forests (Hall and Swaine 1981). Temperature extremes in the forest zone range from 19°C in the coldest month (August) to 33°C in the hottest months (February and March, Hall and Swaine 1981).

## Vegetation and Habitat Diversity

The tropical forest zone of southwestern Ghana has an area of 81,300 km² and constitutes 34% of the country. There are 2,200 recorded forest plant species including many important timber tree species. Seven major forest types have been identified within this zone (Hall and Swaine 1976, 1981). Draw River, Boi-Tano, and Tano Nimiri forest reserves surveyed during this RAP survey belong to the wet evergreen forest type whereas Krokosua Hills forest reserve is at the transition of the moist evergreen to the moist semi-deciduous forest type.

## Description of the RAP Study Sites

We investigated four areas designated as Globally Significant Biodiversity Areas (GSBAs) within four forest reserves. For clarification, in most cases the area of each forest reserve is larger than the area designated as a GSBA. During this survey, we investigated only the area within each forest reserve included within the GSBA delineation. The forest reserves are managed by Ghana's Forestry Commission (FC) under the Biodiversity component of the Natural Resource Management Program (NRMP). The RAP survey took place from 22 October to 10 November 2003, at the end of the rainy season, and covered three sites:

**Draw River:** Wet evergreen forest type
The first RAP survey site (Site 1), Draw River, is situated in the southwestern portion of Ghana's high forest zone, and is contiguous to Ankasa Resource Reserve and Nini Suhien NP (which is west of Draw River). The zone of wet evergreen rainforest averages over 1,750 mm of rain per year (Hall and Swaine 1981). The forest reserve is divided into three blocks by rivers and the Ankrako-Kokum path. There are 40 ha of allowed farms within the reserve. Logging started in 1978 although no trees were removed between 1981 and 1985. In 1988 felling peaked at 712 trees, and figures after 1989 (332 trees) are not available. There is much variation in genetic heat in this reserve depending on soil and landform, though even the "coolest" areas have a high genetic heat (i.e. rich in rare plant species). Draw River ranks 8th of all forests in Ghana with regards to Genetic Heat Index (GHI).

**Boi-Tano (/Tano Nimiri):** Wet evergreen forest type
The second RAP survey site (Site 2) was Boi-Tano/Tano Nimiri, focusing mostly on Boi-Tano. Access in many areas here was difficult due to watercourses and swamps. The GSBA covers an eastern part of the forest in the southern half, including portions along the Tano River, and in the northern section of the forest. Logging of this forest progressed rapidly in the late 1990s, and by 2002 all non-GSBA compartments from Samreboi to the southwestern tip were opened up, forming an extensive logging road network clearly visible on satellite imagery (Hawthorne 2002). The forest here has sample GHIs often in excess of 250. The Tano River and larger tributaries support communities of riverine species with considerable variation between ridge-top and low-lying forests.

**Krokosua Hills:** North-west subtype of Moist Semi-deciduous forest

The third RAP survey site (Site 3) was Krokosua Hills. Our surveys were conducted from the small village of Mmem (population around 75) legally remaining several kilometers within the reserve boundary. The highest point within the reserve is 594 m. Krokosua Hills forest reserve consists of two forest blocks, one above and one below the Asempaneye-Kumasi road. The southern portion of the reserve is heavily hunted and structurally degraded and contains numerous agricultural plots (Magnuson 2002). During the RAP survey, only the northwestern portion of the reserve (corresponding to the GSBA) was surveyed. Because this portion has a relatively small human population and more dramatic topography, the forest here has remained in a relatively more intact state than the southern portion (Magnuson 2002). Krokosua receives 1,250-1,500 mm of rain per year. This range of hills includes high, flat-topped bauxite-capped hills near Ntereso, claimed by locals to be an abandoned mining village. Within the reserve there are a number of farms, old mine pits and villages. Alluvial gold mining was carried out in the southwestern portion in the past (Hawthorne 2002). A botanical survey in 1991 confirmed extreme patchiness with very damaged, logged areas on flatter land and unlogged areas with less damage on steep slopes. The GSBA was last logged in 1988. Chain sawing activity, mainly at night, poses a major problem in the area (and was observed during the RAP). Trees in this type of forest become taller than those of other Ghanaian forest types; heights often exceed 50 to 60 m. Upper canopy consists of deciduous and evergreen species in varying proportions.

### Political Context

In 1957, Ghana became the first country in colonial Africa to gain its independence. Several coups ended in the suspension of the constitution in 1981 and the banning of political parties. In 1992, a new constitution was approved restoring multiparty politics. Lt. Jerry Rawlings, who had acted as head of state in Ghana since 1981, won presidential elections in 1992 and 1996 under the new constitution, but was constitutionally prevented from running for a third term. He was succeeded in 2000 by John Kufuor who defeated the Vice President under Rawlings in a free and fair election. Successive elections, since 1992, reflect the respect for constitutional rule and adherence to the tenets of modern democratic principles leading many within the international community to point towards Ghana as a model of democracy for the rest of the West African sub-region.

### Legal Protection Status

In 1999, the government of Ghana obtained the assistance of the Global Environment Facility (GEF) to implement legal establishment of Globally Significant Biodiversity Areas (GSBAs), designated based on Genetic Heat Index (GHI) of a reserve's botanical species. For the purpose of prioritizing plant conservation in Ghana, each plant species has been assigned to a star category, based on rarity. Black star species are internationally rare and uncommon in Ghana and urgent attention to the conservation of these species is called for. A high GHI signifies that an area is relatively rich in rare, black star species such that loss or degradation of the area would represent a highly significant erosion of genetic resources from the world, and Ghana in particular (Hawthorne and Abu-Juam 1995). In principle no logging or hunting should take place in GSBAs. However, it has never been the mandate of the Forest Services Division of the Forestry Commission to implement hunting regulations and the Forestry Commission does not have the staff to police hunting within the GSBAs. Thus, larger mammals have become increasingly scarce. Whereas much information exists on Ghana's flora, surveys of fauna within these areas are incomplete, outdated, or non-existent.

The mission of the Wildlife Division of Ghana is to work efficiently with others to ensure sustainable management and development of Ghana's wildlife and habitats in order to, "optimize their contributions to national socioeconomic development." This mission immediately indicates the significance of wildlife for commercial use in Ghana (Kormos et al. 2003). The cultural significance and economic benefits of bushmeat in Ghana is staggering and has certainly taken its toll on forest wildlife.

In 1961 Ghana adopted the Wild Animals Preservation Act (Act 43) that regulated export and hunting of "wild animals, birds and fish" in Ghana. This was later strengthened by the list of Wildlife Conservation and Wildlife Reserves Regulations, respectively, introduced in 1971. Ghana became a signatory to the Convention on International Trade in Endangered Species of Wild Fauna and Flora (CITES) in 1976. It was the second country in Africa to do so, following Nigeria.

In 1999 and 2000, wildlife management in Ghana shifted slightly as the once Wildlife Department which was part of the civil service was re-classified as an autonomous division of the Forestry Commission, alongside the Forest Services Division. Some resulting changes to date include much closer collaboration between the Forest Services Division and the Wildlife Division and a relaxation of the laws regarding snail collection in some forest and protected areas. The latter in particular has already had seriously negative effects on wildlife as snail collection is often the gateway through which many people gain access to forest areas for hunting, albeit illegally (Kormos et al. 2003).

Forest reserves controlled by the Forestry Division often have so few staff that they have trouble maintaining boundary lines let alone patrolling for poaching activities. Forest areas controlled by the Wildlife Division enjoy slightly more protection but often are not adequately patrolled, having only minimal effect at reducing hunting activities. Patrol efforts are also not well standardized or regulated and are often inefficient due to the use of wide patrol trails that are easily recognized (and subsequently avoided) by hunters (Kormos et al. 2003). It has been observed that hunting pressure in forest areas often increases dramatically within a few meters of a standard patrol trail (Magnuson 2002).

## Population Profile

In a 2004 census, the population of Ghana was roughly 21,000,000 with a growth rate of 1.25% (CIA World Factbook 2004). There are many different ethnic groups in Ghana speaking over 50 different languages and dialects. The largest ethnic groups are the Akan (44%), Moshi-Dagomba (16%), Ewe (13%), Ga (8%), Gurma (3%) and Yoruba (1%) (Gnielinski 1986, CIA World Factbook 2004). The population of the western region, where the target reserves were located is about 2,000,000. The dominant ethnic group is the Akan.

## Economy

Ghana has roughly twice the per capita income ($450 USD) of poorer West African countries, yet remains heavily dependent on international financial and technical assistance. Major sources of foreign exchange include gold, timber and cocoa, while the domestic economy is heavily reliant on subsistence agriculture. Gross domestic product (GDP) is estimated to be $2,200 USD per capita (purchasing power parity) with 31.4% of the population living below the poverty level (CIA World Factbook 2004).

## Threats to Conservation

Shifting agriculture must have occurred in Ghana centuries, but the rate of deforestation accelerated about a century ago because of timber demands for newly mechanized gold mines, development of communications and a rapidly expanding area of farmland, including cocoa (Hawthorne and Abu Juam 1995).

Ghana has lost roughly 80% of its forest habitat since the 1920s (Cleaver 1992) and about one-third of Ghana's forest is estimated to have disappeared in 17 years between 1955 and 1972 (Hall 1987). Of the original forest zone covering 82,260 km², the area under forest in 1973 amounted to 20,530 km² including 16,790 km² within forest reserves distributed more or less throughout the forest zone (Anon 1973). Forest cover in Ghana was estimated to be 15,842 km² with an annual deforestation rate (1981-5) estimated at 220 km² (Sayer et al. 1992). The most recent figure for "dense forest" based on satellite images is 11,930 km² (Mayaux et al. 2004), which closely confirms the above deforestation rate for the years from 1985 to 2003.

Virtually all forests in reasonable condition today are in fact old designated reserves under the supervision of the Forest Services Division of the Forestry Commission. Many of these forests have retained a significant integrity, in the sense that the boundary lines laid down seventy years ago are still respected and regularly cleared and quite prominent. Various degrees of selective logging have taken place, but in most cases these areas maintain the possibility of regenerating to high forest. There are no real data but a rough estimate would be that about 10-12% of the original forest remains in at least moderate condition, with some pristine blocks in inaccessible areas.

It is estimated that 90% of Ghana's population eats bushmeat when available (Ntiamoa-Baidu 1997). These enormous pressures on wildlife have had drastic implication on several rare species including the chimpanzee (*Pan troglodytes verus*), which is the rarest primate in Ghana (Magnuson et al. 2003) after Miss Waldron's red colobus (*Procolobus badius waldroni*), now officially considered extinct in Ghana (Oates et al. 2000). Recent research also links bushmeat hunting and wildlife decline in Ghana to fluctuating supplies of fish calling for the need to develop cheap protein alternatives to bushmeat and improved fisheries management (Brashares et al. 2004).

## REFERENCES

Anon. 1973. Annual Report of the Forestry Department, Ministry of Lands and Mineral Resources, Ghana, for 1972. Forestry Division, Accra.

Bakarr, M., B. Bailey, D. Byler, R. Ham, S. Olivieri and M. Omland (eds.). 2001. From the Forest to the Sea: Biodiversity Connections from Guinea to Togo. Washington DC: Conservation International.

Beier, P., M. van Drielen and B. O. Kankam. 2002. Avifaunal collapse in West African forest fragments. Conservation Biology, 16:1097-1111.

Brashares, J.S., P. Arcese, M.K. Sam, P.B. Coppolillo, A.R.E. Sinclair and A. Balmford. 2004. Bushmeat Hunting, Wildlife Declines and Fish Supply in West Africa. Science, 306:1180-1183.

CIA. 2004. The World Factbook. Online. Available: http://www.cia.gov/cia/publications/factbook/ August 27 2004.

Cleaver, K. 1992. Deforestation in the western and central African rainforest: the agricultural and demographic causes, and some solutions. Pages 65-78. *In:* Cleaver, K., M. Munasinghe, M. Dyson, N. Egli, A. Penker and F. Wencelius (eds.). Conservation of West and Central African Rainforests. The World Bank/International Union for the Conservation of Nature, Washington, DC. 351 pp.

Davis, S.D., V. H. Heywood and A.C. Hamilton (eds.). 1994. Centers of Plant Diversity: A Guide and Strategy for their Conservation (Volume 1). Gland, Switzerland: World Wide Fund for Nature and IUCN-The World Conservation Union.

Furon, R. 1963. Geology of Africa. (Engl. edition translated by A. Hallam and L. A. Stevens) Edinburgh and London: Oliver & Boyd. xii+377 p.

Gnielinski, S. von. 1986. Ghana: tropisches Entwicklungsland an der Oberguineaküste. Wissenschaftliche Buchgesellschaft, Darmstadt, Germany. xviii+278 pp.

Grubb, P., T.S. Jones, A.G. Davies, E. Edberg, E.D. Starin, J.E. Hill. 1998. Mammals of Ghana, Sierra Leone and the Gambia. The Trendrine Press, Zennor, St. Ives, Cornwall. vi+265p.

Grubb, P. 1982. Refuges and dispersal in the speciation of African forest mammals. Pp. 537-553. *In:* Prance, G.T. (ed.). Biological diversifcation in the tropics. Columbia University Press, New York. xvi+714.

Hamilton, A.C and D. Taylor. 1991. History of climate and forests in tropical Africa during the last 8 million years. Pp. 65-78. *In:* Myers, N. (ed.). Tropical forests and climate. Dordrecht: Kluwer Academic Publishers.

Hall. 1987. Conservation of forest in Ghana. Universitas. 8:33-42. University of Legon, Ghana.

Hall, J.B. and M.D. Swaine. 1976. Classification and ecology of closed-canopy forest in Ghana. Journal of Ecology 64:913-915.

Hall, J.B. and M.D. Swaine. 1981. Distribution and Ecology of vascular plants in a tropical rain forest. Forest vegetation in Ghana. Geobotany 1. Junk, The Hague. 383 pp.

Hawthorne, W.D. 2002. Final report of the floral survey of the Biodiversity Component of NRMP. Forestry Commission, Biodiversity Conservation Component. Ministry of Lands and Forestry, Ghana.

Hawthorne, W.D. and M. Abu-Juam. 1995. Forest Protection in Ghana. IUCN/ODA/Forest Department Republic of Ghana, Gland, Switzerland, and Cambridge, UK, xvii + 203 pp.

Hawthorne W.D. 1996. Holes and the sums of parts in Ghanaian forest: regeneration, scale and sustainable use. Proceedings of the Royal Society of Edinburgh 104B:75-176.

Kormos, R., C. Boesch, M.I. Bakarr and T. Butynski (eds.). 2003. West African Chimpanzees. Status Survey and Conservation Action Plan. IUCN/SSC Primate Specialist Group. IUCN, Gland, Switzerland and Cambridge, UK.

Lebbie, A. 2001. Distribution, Exploitation and Valuation of Non-Timber Forest Products from a Forest Reserve in Sierra Leone. PhD Dissertation, University of Wisconsin-Madison, USA.

Magnuson, L.E. 2002. Distribution and Habitat Use of the Roloway Guenon (*Cercopithecus diana roloway*) in Ghana, West Africa. Master's thesis, Natural Resources: Wildlife Management, Humboldt State University. 68 pp.

Martin, C. 1991. The rainforests of West Africa: ecology - threats - conservation. Birkhäuser Verlag, Basel. 235 pp.

Mayaux, P., E. Bartholomé, S. Fritz and A. Bedward. 2004. A new land-cover map of Africa for the year 2000. Journal of Biogeography 31:861-877. (URL: http://www-gvm.jrc.it/glc2000/publications.htm)

Myers, N. 1982. Forest refuges and conservation in Africa - with some appraisal of survival prospects for tropical moist forests throughout the biome. Biological diversifi-cation in the tropics. G. T. Prance. New York, Columbia University Press: 658-680.

Ministry of Environment and Science. 2002. National Biodiversity Strategy for Ghana. 55 pp.

Ntiamoa-Baidu, Y. 1997. Wildlife and Food Security in Africa. FAO Conservation Guide, 33. Online. Available: http://www.fao.org/docrep/w7540e/w7540e00.htm. August 27 2004.

Oates, J.F., M. Abedi-Lartey, S. McGraw, T.T. Struhsacker and G.H. Whitesides. 2000. Extinction of a West African red colobus monkey. Conservation Biology, 14:1526-1532.

Ojo, O. 1977. The climates of West Africa. Heinemann, London, xvii+218 pp.

Sayer, J.A., C.S. Harcourt and N.M. Collins. 1992. Chapter 21 - Ghana. Pp. 183-192. *In:* Sayer, J.A., C.S. Harcourt and N.M. Collins (eds.). The Conservation atlas of tropical forests - Africa. Macmillan, IUCN, New York.

Thompson, H.N. 1910. Gold Coast: report on forests. Colon. Rep. Miscell. 66:1-238.

White, F. 1983. The vegetation of Africa: A descriptive memoir to accompany the Unesco/AETFAT/UNSO vegetation map of Africa. Paris, Unesco. 356 pp.

A biological assessment of the terrestrial ecosystems of the Draw River, Boi-Tano, Tano Nimiri and Krokosua Hills forest reserves, southwestern Ghana

27

# Chapter 2

Rapid Assessment of Plant Biodiversity
of sections of Draw River, Boi-Tano and
Krokosua Hills Forest Reserves in Ghana

*Patrick Ekpe*

## SUMMARY

This report summarizes the existing knowledge of the vegetation and flora and the results of a rapid assessment of plant biodiversity of sections of Draw River, Boi-Tano and Krokosua Hills forest reserves. The assessment took place from 22 October to 10 November 2003 as part of Conservation International's Rapid Assessment Program (RAP) partly in response to recommendations made for plants during the 1999 Conservation Priority Setting Workshop held in Ghana (Bakarr et al. 2001).

The forest areas investigated form part of Ghana's biodiversity hotspots designated as Globally Significant Biodiversity Areas (GSBAs). Draw River and Boi-Tano fall within the Wet Evergreen forest type and share the same characteristics. Krokosua Hills, which falls within the North-west subtype of Moist Semi-deciduous forest zone, has Upland Evergreen forest type occurring in steep-sided hills. Plant species of global conservation concern have been highlighted. A complete and annotated checklist of plant biodiversity is included.

In Krokosua Hills, the greatest priority for resource managers should be halting the spate of illegal clearance of forest for the cultivation of cocoa and embarking on innovative sustainable livelihood initiatives that are compatible with conservation of biodiversity. Conservation is likely to succeed when the people living close to biodiversity are aware of its value and able to enjoy the benefits from its preservation and sustainable utilization.

## INTRODUCTION

This report outlines the results of the rapid assessment of the plant biodiversity of the sections of Draw River, Boi-Tano and Krokosua Hills forest reserves designated Globally Significant Biodiversity Areas (GSBAs). The assessment took place from 22 October to 10 November 2003 as part of Conservation International's Rapid Assessment Program (RAP) partly in response to recommendations made for plants during the 1999 Conservation Priority Setting Workshop held in Ghana (Bakarr et al. 2001). The three forest reserves form part of reserves designated Globally Significant Biodiversity Areas (GSBAs) managed by the Forest Services Division (FSD) of the Forestry Commission (FC) under the biodiversity component of the Natural Resource Management Programme (NRMP), funded by the Global Environment Facility (GEF).

It is important to state here that the designation and acronym GSBA was previously known as SBPA (Special Biological Protection Area, Hawthorne and Abu Juam 1995, Hawthorne et al. 1998). In 1998 this was changed to GSBA to conform to GEF terminology, when the Ghana Ministry of Lands and Forestry obtained World Bank/GEF funding for the biodiversity component of the NRMP.

The purpose of the vegetation survey during this expedition was to provide additional 'backstop' technical information to lend credence to the conservation value of these areas. Our current survey benefits from previous in-depth work done in this region particularly by Hawthorne et al. (2002), which is the main floral survey report of all of Ghana's GSBAs.

## MATERIALS AND METHODS

Given the fact that the flora of these areas is fairly well-known and the survey was based on a one-person effort, an adaptation of the 'Meander Search' browsing procedure with an 'open eye' for rare and indicated species was adopted. This involves recording species along transects cut for zoological surveys and footpaths with a bias for species known to be endemic to a small part of the globe, i.e. "Black Star" species (Hawthorne and Abu Juam 1995) and also those suspected to have missed the eye/net of the previous survey work. This is done for the purpose of updating the plant checklist of these reserves. To allow a greater chance of sampling across landscape units and vegetation types that would otherwise have been overlooked, attention was paid to riversides, swamps and hilltops. For species that could not be readily identified in the field, specimens were collected for further identification in the Ghana Herbarium.

The vegetation of each study site was scored using a 1-6 point score based on a brief ecological description of the site (Hawthorne and Abu-Juam 1995). This provides a measure for assessing the forest quality in a given forest reserve (Table 2.1).

## RESULTS AND DISCUSSION

Appendix 1 gives a summary of the composition of plant biodiversity in the study area, categorized by family, life-form or habit, ecological guild and conservation rating. A total of 681 vascular plant species belonging to 98 families have been recorded. Life-form categories are summarized in Table 2.2. An overview of some bioquality indices is presented below on individual reserve/GSBA basis.

### Draw River
*Forest Type*

This GSBA falls within the Wet Evergreen forest type (Hall and Swaine 1976). The structure of this forest type is large trees with vertically compressed canopy trees rarely exceeding 40 m. In places, often on slopes, the large shrub *Scaphopetalum amoenum* forms a dense, virtually monospecific understorey. Moist conditions of this forest type support other life-forms like ferns, which are confined to tree boles and branches. This phenomenon is absent or rare in warmer forest types. Characteristic species of this forest type include: *Scaphopetalum amoenum, Cola umbratilis, C. clamydantha, Coula edulis, Cynometra ananta, Pentadesma butyracea, Heri-*

**Table 2.1.** Forest Condition Score

| Score | Definition |
|---|---|
| 1 | EXCELLENT: Few signs (< 2%) of human disturbance (logging/farms) or fire damage, with good canopy and virgin or late secondary forest throughout. |
| 2 | GOOD: < 10% heavily disturbed. Logging damage restricted or light and well dispersed. Fire damage none or peripheral. |
| 3 | SLIGHTLY DEGRADED: Obviously disturbed or degraded and usually patchy, but with good forest predominant, max. 25% with serious scars and poor regeneration; max. 50% slightly disturbed, with broken upper canopy. |
| 4 | MOSTLY DEGRADED: Obviously disturbed and patchy, but with bad forest predominant; 25 - 50% serious scars but max. 75% heavily disrupted canopy. Or forest lightly burnt throughout. |
| 5 | VERY POOR: Forest with coherent canopy < 25% (more than three-quarters disturbed), or more than half the forest with serious scars and poor or no forest regeneration; or almost all heavily burnt with conspicuous *Chromolaena odorata* and other pioneers throughout. Not, however, qualifying as condition 6. |
| 6 | NO SIGNIFICANT FOREST LEFT: Almost all deforested with savannah, plantation or farm etc. < 2% good forest; or 2 - 5% very disturbed forest left; or 5 – 10% left in extremely poor condition e.g. as scattered trees or riverine fragments, remnants with little chance of surviving 10 years. |

**Table 2.2.** Summary of Life-form categories

| GSBAs | Families | Life-form categories / No. of species | | | | Total species |
|---|---|---|---|---|---|---|
| | | Herbaceous species | Lianes/Climbers/ Creepers | Shrub species | Tree species | |
| Draw River | 80 | 50 | 120 | 68 | 232 | **470** |
| Boi-Tano | 74 | 25 | 94 | 60 | 220 | **399** |
| Krokosua Hills | 78 | 32 | 81 | 51 | 214 | **378** |

*tiera utilis* and *Chytranthus cauliflorus*. The forest is termed evergreen because of the relative scarcity of deciduous trees in the canopy.

### Forest Condition/Quality

The pattern of vegetation and forest quality revealed that Draw River forest reserve is still in relatively healthy condition with an average score of 2. Old log dumps and hauling tracks, relics of the most recent logging in 1990 are at various stages of recovery. These areas, dominated by pioneer species such as *Trema orientalis, Musanga cecropioides, Cercropia peltata* and *Solanum enriantum,* score 4. Human interference through small scale harvesting of minor forest produce – building poles, canes, raphia palm, trapping for game, etc. – are noticeable in some areas, particularly along the historically important Ankrako-Kokum footpath.

### Conservation value

Eighteen 'Black Star' species are known from Draw River to date (Hawthorne 2002). Nine of these were recorded during the course of this survey, namely: *Alsodiopsis chippii, Berlinia occidentalis, Chytranthus verecundus, Cola umbratilis, Dichapetalum filicaule, Hymenocoleus multinervis, Leptoderris miegei, Placodiscus bancoensis* and *Tapura ivorensis.* The Star rating system adopted by Ghana defines the conservation merit of each forest species in Ghana (Hawthorne 1992, 1996; Hawthorne and Abu-Juam 1995). The star of a species defines its weight for the calculation of weighted average referred to as Genetic Heat Index (GHI), which provides a framework for defining the conservation merit of a tract or sample of forest of any size (Table 2.3). Draw River has performed creditably on the 'National league' of Genetic Heat Index (GHI) by placing eighth (Hawthorne and Abu-Juam 1995).

### Boi-Tano
#### Forest Type

Boi-Tano, like Draw River, falls within the Wet Evergreen forest type and shares the same characteristics.

### Forest Condition/Quality

An average score of 2 is recorded for the GSBA area. An extensive network of logging roads has opened up the reserve creating more access than in the early 1990s. Non-GSBA compartments are dotted with logging scars and dominated by pioneer species. More severely disturbed areas are log dumps and road verges, scoring 4 on the forest condition assessment chart. The main hauling road appears unusually wide. Its verges, on either side, present a picturesque image of a *Cercropia peltata* plantation, an invasive pioneer tree species recorded for the first time along the Elubo-Côte d'Ivoire road in the early 1990s and now seen as far inland as Ayum Forest Reserve in the Goaso District.

### Conservation value

This GSBA, which occupies the fifth position on the league of GHIs, continues to make interesting revelations by way of rare plants records. Following this survey, the number of black star species known from this reserve is 23, 13 of which were recorded during this survey. These are *Alsodiopsis chippii, Berlinia occidentalis, Chytranthus verecundus, Cola umbratilis, Dichapetalum filicaule, Landolphia membranacea, Leptoderris miegei, Momordica sylvatica, Nepthytis swainei, Placodiscus bancoensis, Schumanniophyton problematicum, Tapura ivorensis* and *Tetrapleura chevalieri.* Notable among these species are *Momordica sylvatica,* recently described (Jongkind 2002) based on specimens from Côte d'Ivoire, and *Tetrapleura chevalieri.* To date, both species are known

**Table 2.3** Summary of star categories of conservation priority for species.

| Star | No. of Species | No. of Degree squares occupied | Weight for GHI | Note |
|---|---|---|---|---|
| Black | 52 | 1.0 ± 0.5 | 27 | Highly significant in context of global biodiversity; Rare globally and not widespread in Ghana. |
| Gold | 208 | 7.8 ± 3.8 | 9 | Significant in context of global biodiversity; fairly rare globally and/or nationally. |
| Blue | 414 | 24.5 ± 12.6 | 3 | Mainly of national biodiversity interest; e.g. globally widespread, nationally rare; or globally rare but of low concern in Ghana due to commonness. |
| Scarlet | 14 | (included in Red) | 1 | Common and widespread commercial species; potentially seriously threatened by overexploitation. |
| Red | 40 | 39.6 ± 16 | 1 | Common and widespread commercial species; under significant pressure from exploitation. |
| Pink | 19 | (included in Red) | 1 | Common and widespread commercial species; not currently under significant pressure from exploitation. |
| Green | 1044 | 69.2 ± 49.8 | 0 | Species common and widespread in tropical Africa; no conservation concern. |
| Other | > 100 | | | Unknown, or non-forest species e.g. ornamentals or savannah plants. |

in Ghana only from this forest reserve. Ntim Gyakari discovered the latter in Ghana for the first time in 1998. It could be argued that the former species was recorded for the first time in Ghana during this survey, even though photographs of fruits of an unidentified plant taken by Hawthorne and Ntim Gyakari in this forest reserve in 2001, prior to the description of the species, was later found out to be the fruits of *Momordica sylvatica*.

## Krokosua Hills

### Forest Type

Krokosua Hills falls within the northwest subtype of Moist Semi-deciduous forest. Trees in this forest type become taller than those of other forest types; heights often exceed 50 m and sometimes 60 m. The upper canopy consists of deciduous and evergreen species in varying proportions. Characteristic species include *Argomuellera macrophylla*, *Chlamydocarya thomsoniana*, *Guibourtia ehie*, *Periscopsis elata* and *Khaya anthotheca*. However, Upland Evergreen forest type occurs in steep-sided hills (500-750 m elevation) in this forest reserve. Structurally, the tallest trees rarely exceed 45 m and great diversity of epiphyte species and life-forms is characteristic.

The presence of plant species associated with Southern Marginal forest types – *Christiana africana*, *Elaephorbia drupifera*, *Mallotus oppositifolius*, *Strophanthus hispidus* and *Hildegardia barteri* – near GSBA Boundary Pillar 72 along the Mmem-Sikaniasen footpath and a few other places is indicative of shallow soils e.g. around granite domes.

### Forest Condition/Quality

The forest in most parts of the GSBA is slightly degraded, but relatively good forest characteristic of this forest type predominates, i.e. condition 3 on the forest condition chart. Portions east of Mmem, a village legally allowed within the GSBA, are obviously degraded. The absence of an upper canopy and non-coherent middle canopy and climber-tangled understorey are common sights. This portion is recovering from bush fire, which swept through the forest in 1993.

Relics of chain-sawed lumber are a common phenomenon. The disheartening attack on the GSBA is the illegal clearing of the best part of the forest for farming purposes by inhabitants of Mmem village. Focal Area 2, a research facility established in the course of the floral survey of the GSBA, was completely destroyed and planted with cocoa seedlings and plantain. Nine of such pockets of illegal farms, ranging between 0.5 ha and 1.5 ha and all less than two years old, were encountered during this survey. Investigations revealed that people from fringe communities are now moving into Mmem where they have been allocated forest land for farming under a share-cropping agreement with the chief.

Forest Services Division staff attempted replanting the illegal farms using *Cedrela odorata*, an exotic species, which we suggest is inappropriate within a GSBA. Any planting scheme in a GSBA should use species indigenous to Krokosua Hills. The planting exercise in Krokosua Hills on the whole is too cosmetic to make any impact as only the peripheries of the farms are planted. The tree seedlings are planted among the young cocoa seedlings without first uprooting the cocoa. Most of the *Cedrela odorata* seedlings are dying (as the planting material is too weak to survive) whereas the cocoa seedlings are doing well. A few potted *Mansonia altissima* (an indigenous species) were seen dumped at the edge of one the farms. This planting material is too young and weak and cannot survive when planted.

A kilometer radius of forest has been demarcated awaiting clearing as an extension of farmland for the village. This demarcation was said to have been done with the chief's authority, when his request to FSD Headquarters in Accra for extension of farmland for the village was not granted. Given the current pace of forest clearance for agriculture purposes, the whole GSBA may be converted to cocoa farms within the next ten years. The reserve of bauxite, although of relatively low grade (Cooper 1958), suggests another potential threat to the Upland Evergreen portions of Krokosua Hills GSBA.

### Conservation value

Three Black star species, *Cassipourea hiotou*, *Coffea togoensis* and *Dasylepis assinensis* and a fairly high number of Gold star species are found here. All three black star species were recorded during this survey in addition to *Hymenostegia aubrevellei*, a gold star species that, in Ghana, is known only from this GSBA.

Krokosua Hills was not originally proposed as a GSBA (it had been a Hill Sanctuary), due to its relatively low GHI values and patchiness, but was later added because of the recent presence of Miss Waldron's red colobus *(Colobus badius waldroni)*, now considered extinct (Oates et al. 2000). This statement needs to be confirmed by further more extensive fieldwork.

## CONSERVATION RECOMMENDATIONS

- Establish a special task force, perhaps with military support, as a matter of urgency, to save Krokosua Hills GSBA. The FSD should not lose sight of the fact that Sukusuku, Manzan and Bodi forest reserves were lost under similar circumstances.

- Conduct a stakeholders' forum, led by the Forestry Commission, to declare the status of Krokosua Hills and to determine who is in charge of the day-to-day management. Forest Services Division personnel appeared to have been replaced by Wildlife Division staff.

- Conduct a survey of the internal boundary at Mmem village as soon as possible. Subsistence and cocoa farming at Mmem continues to encroach upon the forest.

- Promote the market for specific NTFPs to shift the emphasis from destructive uses of GSBAs for non-sustainable economic gains. Explore the market potential of Non-Timber Forest Products (NTFPs) like *Thauma-*

*tococcus daniellii* yielding Thaumatin, a substitute for sugar, ideal for diabetic patients (Hedberg 1986), and vegetable oil from the nuts of *Allanblackia parviflora* (Abbiw 1990). *Allanblackia parviflora* holds potential to provide a source of household income for rural populations while contributing to forest landscape restoration. IUCN and Unilever are partners in Novella Africa, a pilot project that focuses on the sustainable utilization of *Allanblackia parviflora* seeds. Both parties seek to promote socially acceptable and environmentally sound market-based financing mechanisms to safeguard both biodiversity and livelihoods in the tropical belt of Africa. Towards the fulfillment of this aim, IUCN is preparing guidelines that will be used by harvesters, local people in the villages, and buyers in the *Allanblackia parviflora* market chain.

In conclusion, destruction of forest biodiversity by people living around protected areas casts doubts on the emerging 'dogma' that recognizes that people in and around protected areas can be stewards of forest biodiversity. More work needs to be done on this principle in relation to securing local rights and equitable benefit-sharing mechanisms to ensure that livelihoods thrive alongside managed biodiversity. Conservation is likely to succeed when the people living close to biodiversity are aware of its value and able to enjoy the benefits from its preservation and sustainable utilization.

## REFERENCES

Abbiw, D.K. 1990. Useful plants of Ghana. Intermediate Technology Publications Ltd. and The Royal Botanical Garden, Kew, London, xii + 337 pp.

Bakarr, M., B. Bailey, D. Byler, R. Ham, S. Olivieri and M. Omland. 2001. From the forest to the sea: biodiversity connections from Guinea to Togo. Conservation Priority-Setting Workshop, December 1999. Conservation International, Washington, DC, 78 pp.

Cooper, W.G.G. 1958. The bauxite deposits of the Gold Coast. Bull. Geol. Survey of the Gold Coast, 7: 1- 33.

Hall, J.B. and M.D. Swaine. 1976. Classification and ecology of closed-canopy forest in Ghana. Journal of Ecology, 64:913-915.

Hawthorne, W.D. 1990. Field Guide to the forest tree of Ghana. Chatham: Natural Resources Inst. Ghana Forestry Series 1, vi + 278pp.

Hawthorne, W.D. 1992. Forestry, dragons and genetic heat. Paper presented at seminar on conservation aim in Africa. Wildlife Conservation International. June 1992.

Hawthorne, W.D. 1995. Ecological profiles of Ghanaian forest trees. Tropical Forestry Paper 29. Oxford Forestry Institute. vi + 345pp.

Hawthorne, W.D. 1996. Holes and the sums of parts in Ghanaian forest: regeneration, scale and sustainable use. Proceedings of the Royal Society of Edinburgh, 104B:75-176.

Hawthorne, W.D. 2001. Forest Conservation in Ghana: forestry, dragons, genetic heat. Pp. 491-512. *In:* Webber, W., L.J.T. White, A. Vedder and L. Naughton-Treves (eds.). African rain forest ecology and conservation: an interdisciplinary perspective. Yale University Press, New Haven and London, xii + 588 pp.

Hawthorne, W.D. 2002. Final report of the floral survey of the Biodiversity Component of the NRMP. Forestry Commission, Biodiversity Conservation Component. Ministry of Lands and Forestry, Ghana.

Hawthorne, W.D. and M. Abu-Juam. 1995. Forest Protection in Ghana. IUCN/ODA/Forest Department Republic of Ghana, Gland, Switzerland, and Cambridge, UK, xvii + 203 pp.

Hawthorne, W.D., M. Grut and M. Abu-Juam. 1998. Forest production and biodiversity conservation in Ghana, and proposed international support of biodiversity conservation. CSERGE working paper, GEC 98 - 18.

Hedberg, I. 1986. Conservation and land use. Pp. 309-326. *In*: Lawson, G.W. (ed.). Plant ecology in West Africa. Systems and processes. John Wiley and Sons Ltd., Chichester, New York.

Jongkind, C. 2002. A new species of *Momordica* from West Africa. Blumea, 47: 343- 345.

Oates, J.F., M. Abedi-Lartey, S. McGraw, T.T. Struhsacker and G.H. Whitesides. 2000. Extinction of a West African red colobus monkey. Conservation Biology, 14:1526-1532.

# Chapter 3

## Rapid Assessment of Butterflies of Draw River, Boi-Tano and Krokosua Hills

*Torben B. Larsen*

## SUMMARY

The three forest areas visited during our survey have been designated as Globally Significant Biodiversity Areas (GSBAs), the concept of which is a welcome one since they increase the protected areas in Ghana and provide 'stepping stones' between some of the three existing small national parks in the forest zone. Butterflies are good ecological, biodiversity, and biogeographical indicators and the time allocated for the RAP mission was more or less adequate, though a further round during another season would have been desirable. Unfortunately, an excessive and extended rainy season and poor weather conditions during the mission made butterfly collecting disappointing. In all, 250 species were collected, well below the 400 or so, mostly forest species, indicated as possible through prior experience in Ghana. Nonetheless, one species was new to Ghana, and several rare or otherwise interesting species were recorded. The butterflies imply that: 1) Draw River forest reserve is a forest in good condition for butterfly biodiversity and with a typical butterfly population characteristic of the wet evergreen forests. At present only Ankasa/Nini Suhien protects this type of forest. Since Draw River is contiguous with Ankasa, it seems logical to merge the two; 2) Boi-Tano forest reserve is another forest in good condition that warrants increased protection. Hunting pressure here is extreme. It is a valuable forest in its own right and, as it is close to Nini Suhien, can act as a 'stepping stone' for gene-flow between this and Bia. 3) The Krokosua Hills forests are in large part in worse shape than the two previous. There is considerable encroachment of cocoa, hunting is extreme, there is/was some illegal logging, and official logging has been less sensitive. The encroachment process may be irreversible. However, the central ridge has intact forest, which produced a rare butterfly new to Ghana and another surprising record and many species not seen elsewhere during the RAP survey. The fauna of the central ridge is evidently much more important than the degraded lower areas. It could possibly be protected by being nominated as a core area as a trade-off for a continued 'laisser-faire' policy in the surrounding lower forests.

## INTRODUCTION

About 20,000 species of butterflies (Lepidoptera, Rhopalocera (Papilionoidea and Hesperioidea)) are known worldwide. Broadly speaking they are the best-studied major group of arthropods. Most species have been described and their geographic distributions are relatively well known. Usually some information of their habitat choice, habits, and host plants is available, if not at species level then at least at genus level. They are therefore well suited as ecological, biodiversity, and biogeographical indicators. Among the million arthropods described, butterflies are thus among the best suited for inclusion in missions such as a rapid biological assessment for the purpose of setting conservation priorities. However, it must not be forgotten that different organisms have different temporal and geographical origins. They may react differently to changes. Generalizations cannot be made on the basis of any one group. For example, butterflies are much more robust than soil protozoa; fish are more influenced by watersheds, and so on (Lawton et al. 1998).

About 4,000 butterfly species are Afrotropical (20% of the world fauna). The vast bulk of these are endemic to the African continent and Southwest Arabia. No more than 20 or 30 Afrotropical species can be found outside of Africa/SW Arabia and just a handful of Palaearctic species penetrate into the Afrotropical Region, chiefly in Arabia and Ethiopia. See Larsen (2005) for a discussion of West African butterfly ecology and biogeography.

There are significant differences between regions within Africa mediated by ecological conditions and through past biogeographical developments. For butterflies – and indeed for many arthropods – the most important may be the division between the species of the forest zone and those of open formations (savannah and veldt). Few species, and only a minority of genera, manage to straddle both biomes. The presence of savannah species in forest areas generally denotes that significant environmental degradation has taken place.

There are also some major and many minor biogeographical regions that can be identified through butterfly distribution patterns, several of which are relevant to the fauna of Ghana. These patterns should assist in directing conservation decisions.

Thus, in Ghana, two local areas have a fauna and flora that contains endemic elements found nowhere else. These are the upland forests of the Atewa Range and the Tano Ofin peak, as well as the mountains of the Ghana/Togo border zone (the Volta Region), which are among the most urgent conservation priorities in the country.

At the larger level, the forest zone of Africa west of the so-called Dahomey Gap can clearly be divided into to a Ghana Subregion covering Ghana (except for the Volta Region) and eastern Côte d'Ivoire, and a Liberia Subregion from the Basse Casamance in Senegal to western Côte d'Ivoire. In turn, the two subregions jointly form a wider West African region of endemism (often known as the Upper Guinea Region). An even wider region of endemism extends from West Africa to the Nigeria/Cameroon border regions.

The three forests covered by the present RAP exercise are in the core area of the Ghana Subregion and their conservation value lies in the extent to which they conserve the special species of the subregion, as well as the overall biodiversity of which these species are an important component.

Ghana has a total butterfly fauna of just over 900 species (Appendix 2), almost a quarter of the total Afrotropical fauna. Some of these are pure savannah species, some limited to very dry forests, and others in Ghana are limited to the Volta Region. The number of species that could be present in the areas visited by the RAP should be somewhere around 750. About one in six of these species are among the endemic groups mentioned above.

## MATERIALS AND METHODS

This study is based on butterfly observation and collection by the standard methods in the three areas visited. No previous butterfly records are known from Draw River or from Boi-Tano, but data from two days of collecting at Krokosua Hills in November 2000 by this author are included in the report.

The author has also collected extensively in forests near to the ones studied, especially Ankasa and Bia National Parks. The results of these collections are included in Appendix 2.

Butterflies were collected chiefly with a standard butterfly net while walking in suitable habitats. Only rarely is any individual spot in the forest (a clearing or a stretch of river where butterflies mud-puddle) worth surveying for any longer time period. Normally two kilometres can be covered during an hour of walking while carefully searching for butterflies. The author benefited from the advantage of his prior experience in Ghana, allowing for the identification of some 85% of all species by sight. Under less familiar conditions 50% or more may need to be collected for certain identification. Four traps baited with fermenting fruit, rotten prawns, and the carcasses of small mammals were also employed. Such traps often bring in species that are rarely captured with the net, primarily members of the subfamily Charaxinae. More than a dozen rare species were collected only in the traps.

The number of species recorded from any given locality is usually determined by the number of kilometres walked and sampled in fair weather at various times of the year when butterflies are plentiful. I have attempted to standardize my own collecting experience in West Africa as the results of 'collecting days' defined as: "Five hours of concentrated collecting in fair to good weather conditions." A really perfect full day would count as 1.3 'collecting days,' whereas two days with indifferent weather would count as just 1.0 full 'collecting day.' In the forest zone it may rain and/or be heavily overcast in which case hardly any butterflies are found and the day not rated as a 'collecting day' at all. During the RAP not one single day constituted a full 'collecting day' and two were complete washouts.

In the West African forest zone the usual number of species observed during the first 'collecting day' would be 120 or so. The next full day about 50 or 60 additional species would be added, 30 on the third, then 15, etc. Thus, a good week in a given forest should yield about 250 species and further work will then produce a few additional species. A week's collecting at another time of the year will yield similar quantitative results but many of the species will be different; this is because many butterflies are seasonal; only about 60% of the species are on the wing at any given time. After a total of 400 species has been reached in a given forest, additional species accrue very slowly. The total in a forest of reasonable size and in reasonable condition, including species from secondary growth and the forest fringes, should be 600 or more (Larsen 2001), but it takes years of work to reach even 500.

The ideal requirement for a rapid butterfly survey in a new forest would thus be two well-spaced visits of one week under good weather conditions. This was obviously not possible during the present survey, but interpretation of the results from this survey are assisted by the author's familiarity with the butterflies of the area in general.

## RESULTS

Unfortunately, the season was poor for butterflies throughout the RAP mission. The rainy season had been extended and extensive, with the result that especially the butterflies of the dense forest and its floor were at very low density and diversity. The Lycaenidae and the Hesperiidae, in particular, were conspicuous by their relative absence (they form about 60% of the total butterfly fauna). The large forest floor butterflies (genera such as *Euphaedra*, *Euriphene*, and *Bebearia*) were usually worn, sometimes to the point where identification was difficult, and more females were collected than is usual. This indicates little recent hatching – it would seem that most were awaiting improved weather in late larval or pupal stage in a state of quiescence. At Krokosua Hills, the driest locality, visited last, there were more signs of incipient hatching, including several fresh *Euphaedra* species not seen elsewhere.

In addition, the weather was very poor for butterflies, with heavy morning mists or rain, and just a few hours of sunshine before afternoon clouds or rain closed in. Only a few days qualified as more than half a full 'collecting day' as defined above.

### Sites visited

#### a) Draw River

A total of 135 species were collected during 2.5 'collecting days'. Weather was frightful. The forest is contiguous with Ankasa National Park at its western limit and is botanically and visually similar. Ankasa has a large number of plants endemic to the wettest parts of the Ghana subregions as well as many species that are internationally and/or regionally very rare. Botanically Ankasa is one of the places in Ghana with the highest 'genetic heat index' and Draw River is fairly hot (Table 3.1).

The butterfly fauna of Ankasa contains many endemics to the Ghana subregion and to West Africa and has a distinctly 'wet' aspect. There are also a large number of rare species, especially in the family Hesperiidae. There are about 600 species (Larsen 2001); this is slightly less than in other Ghana forests classified as Moist Evergreen or Moist Semi-deciduous due to the absence of many species of the drier forest habitats.

Draw River produced a number of indicator species of the wetter forest systems in good condition (Dall'asta et al. 1994). A good example is *Papilio horribilis*, one of the largest and showiest of the swallowtails. Also present was *Euriphene veronica*, previously recorded from Ghana only in Ankasa. This is basically a Liberian subregion endemic, a vicariant of *Euriphene barombina* that just extends to western Ghana. Only one was found in Draw River together with large numbers of *E. barombina*; in Ankasa *E. veronica* is by far the more common of the two. The two are sympatric only in scattered localities between Draw River and Abidjan. Another interesting species is *Tetrarhanis baralingam*, a rare species that I recently described and believed to be a Ghana endemic, but which has also been found in Sierra Leone.

Though the Hesperiidae were almost wholly absent, a single perfect female of the rare *Pteroteinon pruna* was found, probably an indicator that many of the rare species known from Ankasa are present at Draw River.

#### b) Boi-Tano

A total of 99 species were collected during 2.5 'collecting days.' This is exceptionally low. In addition to bad weather, collecting was hampered by the fact that the roads through the forest were much too wide for good collecting, while there was an absence of the broad walking tracks through the forest which make collecting easy. The various poacher trails were too narrow to concentrate butterflies in any numbers.

A number of positive indicator species for wetter rainforests in good conditions were present even in the limited material recorded. Among the swallowtails, the large and pretty *Graphium illyris* and *G. latreillianus* flew along roads in some numbers, and even came mud-puddling, despite the poor weather. The rare hesperiid, *Hypoleucis sophia*, was also present; there are probably less than 20 of these in collections anywhere.

**Table 3.1** Some data on the three forests visited during the RAP survey

| Name/District | Forest type* | Quality** | Size | Genetic Heat *** |
|---|---|---|---|---|
| DRAW RIVER, Tarkwa | WEF | 2 | 235 km² | 186 Fairly Hot |
| BOI-TANO, Enchi | WEF | 2 | 129 km² | 202 Very Hot |
| KROKOSUA HILLS, Juaboso | MSF | 3 | 482 km² | 46 Tepid |

\*      WEF = wet evergreen forest, MSF = moist semi-deciduous forest
\*\*    2 = good condition 3 = reasonable condition (see plant chapter for further description)
\*\*\*  an index summarizing endemicity and rarity of the flora
Source: Hawthorne and Abu-Juam 1995

A biological assessment of the terrestrial ecosystems of the Draw River, Boi-Tano, Tano Nimiri and Krokosua Hills forest reserves, southwestern Ghana

35

### c) Krokosua Hills

A total of 182 species were collected during 4.0 'collecting days' – previously 120 species were found during two full 'collecting days,' also a suboptimal result (Larsen 2001). Of the species collected earlier, 20 were not seen during the RAP mission, so the total known species from Krokosua Hills is now 202.

Much of Krokosua Hills is heavily logged and many access-roads are very wide and clogged with the invasive weed *Chromolaena odorata*. This weed is hard to manage because of its ability to spread rapidly. It also dries during the dry season, and then allows fire to penetrate into the forest. The village of Mmem where we camped is located in the middle of the forest with permission from the Forest Services Division. Cocoa was the main crop. There was clear recent encroachment on the existing forest, and this is apparently the case also on some of the boundary lines. Negative indicator species for forest in good condition were more plentiful in and around Krokosua Hills than in the other forests (*Colotis euippe, Eurema brigitta*; Dall'asta et al. 1994).

There are still areas of good forest in Krokosua Hills. In an earlier report on Krokosua Hills I stated: "It seems certain that the unlogged areas at higher levels than it was possible to reach on a brief survey will have a more substantial butterfly fauna that is likely to be diverse and interesting. Hilly country often throws up surprises" (Larsen 2001). This prediction was validated through walks traversing the summit of the Krokosua Ridge. Two great surprises were found on this trail:

1) *Euriphene leonis,* essentially a species of the Liberian subregion, rare in Sierra Leone, Liberia, and very scarce in the Taï Forest in Côte d'Ivoire. A few years ago it was surprisingly recorded from Yapo Forest near Abidjan. This is the first Ghana record and I would have rather expected its presence at Ankasa.

2) *Catuna niji,* a species mainly from the Liberian subregion and in Ghana known only from one part of Ankasa, where it is common. I have searched assiduously for it in other forests in Ghana without success.

3) One of the traps also included my first Ghana specimen of the rare *Charaxes hildebrandti*, known from very few Ghana specimens.

Krokosua represents the moist semi-deciduous forests that have lower rainfall than the two others, as evidenced by several species not found at Ankasa, Draw River, or Boi-Tano (e.g. *Dixeia cebron, D. capricornus* and *Junonia westermanni*). During my 2001 visit I found a species tentatively named *Triclema krokosua*; I have not been able to do the final description since it belongs to a difficult complex and insufficient comparative material is on hand. However, it is almost certainly not endemic to Krokosua.

### Endemicity

About 150 of Ghana's butterflies (17% of the 900 species) are West African endemics, though not all of these are present in the area studied (some are limited to the Volta Region or to the Atewa Range). They fall into three categories:

1) Those that are strictly limited to the Ghana Subregion, covering the area between the Volta Region and central Côte d'Ivoire (GHE in Appendix 2);

2) Those that are West Africa endemics, found in both the Ghana and the Liberian Subregions, but not extending further east (WAE); and

3) Endemics that are limited to the area of West Africa and Nigeria, including the extreme west of Cameroon (WNE). These are annotated in Appendix 2. The numbers by category in Ghana and those collected during the RAP are given in Table 3.2.

The proportion of collected endemics in the three forests is rather low (14% of the 250 species recorded). However, many are rare and/or local and less likely to be collected during short visits under poor collecting conditions. This is partly evidenced by the fact that no fewer than 27 endemics were only collected in a single of the three localities. On balance it seems likely that the areas surveyed will contain a significant proportion of the endemic forest species.

**Table 3.2** Summary of butterfly endemicity in Ghana and during the RAP survey

| Type of endemic* | GHANA | DRR | BOT | KRO | RAP |
|---|---|---|---|---|---|
| Ghana Subregion Endemics (GHE) | 25 | 1 | 0 | 0 | 1 |
| West Africa Endemics (WAE) | 61 | 11 | 3 | 13 | 20** |
| WA to Nigeria Endemics (WNE) | 63 | 7 | 4 | 11 | 16 |
| Total Endemics | 149 | 19 | 7 | 24 | **37** |

\* note in Appendix 2

\*\* Two species should probably be considered extreme extensions of species from the Liberian Subregion

DRR = Draw River Forest Reserve, BOT = Boi-Tano Forest Reserve, KRO = Krokosua Hills Forest Reserve,

RAP = all species collected during the rapid assessment survey

*Total*

A combined total of 250 species were collected in the three forests during the RAP survey. This is almost exactly one-third of the total expected in the area covered. Normally, with better weather conditions, I would have expected about 400 species in the time available in the three forests. It is the rare species that are missing from the material. The low numbers in all three localities have not permitted more detailed comparative analysis of the three forest's butterfly faunae or comparison with the totals for Ankasa and Bia. However, balanced by experience from Ankasa and Bia National Parks, the general impression is that Draw River and Boi-Tano are effective reservoirs of the West African wet forest fauna. The hilly parts of Krokosua have a varied and interesting fauna, evidently with some surprises, such as is often the case in hilly country.

## DISCUSSION

As far back as the 1920s and 1930s, farsighted foresters in Ghana realized that most forests would simply disappear unless they were demarcated and placed under some sort of restrictions. Virtually all forests in reasonable condition today are in fact these old designated reserves which are under the supervision of the Forest Services Division. Their main purpose was to ensure the sustainable use of Ghana's forest resources and the preservation of forests with important roles as watersheds and windbreaks. The forest status as of 1985 is given in Table 3.3 below.

Two main conclusions emerge from this table. First, only about 20% of the original rainforest area is still classified as forest. This, however, is an overly optimistic estimate. Much of this forest is in very poor condition, probably too poor for natural regeneration. A significant proportion no longer qualifies as forest in any meaningful sense since the canopy has been wholly destroyed; a more or less complete canopy is necessary to maintain the microclimate to which rainforest organisms have adapted. Some of the area has been converted to plantations, such as teak (*Tectona grandis*), which are effectively biodiversity deserts.

**Table 3.3** Original forest zone, present forest zone, and rainforest area under National Park or Resource Reserve Status

| Forest and National Parks | Area km² | Percent |
|---|---|---|
| Original rainforest area | 82,000 | 100.0 |
| Classified as forest 1985 | 18,000 | 22.0 |
| Total Rainforest National Parks | 970 | |
| •    Ankasa Resource Reserve | 343 | |
| •    Nini-Suhien N.P. | 166 | 1.2 |
| •    Bia N.P.* | 77 | |
| •    Kakum N.P. | 370 | |

\* Bia Resource Reserve is considered too degraded to be classified as forest for the purpose of this survey.

Source: Larsen (2001)

However, many of the forests managed by the Forest Services Division have retained a significant integrity, in the sense that the boundary lines laid down seventy years ago are still respected and the boundary lines regularly cleared and quite prominent. Various degrees of selective logging have taken place, sometimes excessive, but these areas are still forests and they could potentially regenerate to, at least, a semblance of primary forest. There are no real data but a rough estimate would be that about 10-12% of the original forest remains forest in at least moderate condition, with some pristine blocks in inaccessible areas.

On the whole, the Forestry Department has never put much effort into implementing hunting regulations, so in most of Ghana larger mammals have become increasingly scarce. However, the impact on the small members of the fauna such as arthropods has been much less. Conversely, the extensive deforestation has led to an increase in the number of savannah and adventive butterflies. Most of these were probably always present in very localized special habitats, such as rocky areas with no tree-cover within the rainforest zone, and this increase is most noticeable in the drier forests. Isolated inselbergs have even less mobile savannah organisms than butterflies. However, there is no doubt that the numbers of species have greatly changed. The adventive species normally never manage to establish permanent populations inside actual forest in good condition; these species would very rarely establish a competitive relationship with forest species, but are opportunistic users of newly created habitat types. The presence of *Hamanumida daedalus* in Boi-Tano and Krokosua Hills and especially of *Eurema brigitta* in Krokosua Hills are typical examples.

Only about 1.25% of the original forest area is managed as National Parks (Ankasa/Nini Suhien, Bia North, and Kakum). These are managed by the Wildlife Division, which has its own protection staff that – though insufficient in number – is proportionately larger than that of the Forestry Department. There is no doubt that the establishment of the national parks has helped to protect larger animals. However, 1.25% of the original forest area is a very small proportion indeed.

The forests investigated during the RAP mission were managed by the Forestry Department as productive forests, but then reclassified as Globally Significant Biodiversity Areas (GSBAs) during the 1990s. In principle no logging or hunting should take place in GSBAs and the reclassification is to be welcomed as a potentially significant contribution to the maintenance of biodiversity in Ghana. However, under the present conditions the Forest Services Division make available [and train and authorize] the staff to police the GSBAs and the entire emphasis of this division has always, and very naturally, been on the economics of forest production. The world's remaining biodiversity is the heritage of all of humanity and all of humanity has the responsibility for supporting its conservation. Some way must be found to support Ghana in its wise decision to create GSBAs.

If the above estimate that only 10-12% of the original forest area still exists is correct, it might be said that all existing forest ought to be placed under protection; but this would not be realistic. However, it is worth considering the size of the present national parks. Ankasa/Nini Suhien is 509 km², Bia North is 77 km², and Kakum is 370 km². While I suspect that forests such as Nini-Suhien and Kakum are large enough to maintain genetically viable populations of butterflies and most arthropods, this might not be so for the largest of their inhabitants, the elephant. All the national parks at present have a potential gene-flow with some neighbouring forests in reasonable condition. The GSBAs have the potential for providing permanent 'stepping stones' for gene-flow also if the remaining forests continue to diminish in size, which cannot be precluded in the face of continued rapid population growth (in the expectation that Ghana's battle against AIDS will be successful).

The direct effects of deforestation on butterfly diversity can be seen today in most parts of Ghana, but perhaps with the greatest clarity in and around Kakum National Park, Ghana's flagship park for tourism development. Kakum is the best-studied single forest in terms of butterflies and more than 500 species have been recorded from the park, including some degraded areas. A projection of the logistical 'collecting curve' indicates a total of about 630 butterfly species (almost twice that of western Europe and more than twice that of any state in the USA. Texas, with 290 species, being the richest). A butterfly inventory within walking distance of the Hans Cottage Botel near Kakum over many years has only reached 75 species. Some of these are savannah species, never seen at Kakum, most are species of disturbed habitats, and a few are particularly hardy forest butterflies that just manage to survive in a small, scraggly sacred grove in the area. Conversely, I have recorded more than 250 species from the tiny, isolated Boabeng-Fiema Forest (maintained as a sacred grove by the local community, with support from by the Wildlife Division). The 'collecting curve' is still far from flat, so the true total should be at least 300-350. The area was at the northern limits of the forest zone, where some 500 species might have been expected in an intact, unbroken forest zone. This little forest may maintain two-thirds of the original butterfly biodiversity, despite being surrounded by a broad belt of derived Guinea savannah and agricultural lands with less than a hundred mainly common and widespread species.

## CONSERVATION RECOMMENDATIONS

The designation of GSBAs is to be welcomed. New protected areas are to be welcomed not only in their own right, but also as 'stepping stones' to facilitate future gene-flow. Most of the small sacred groves visited in Ghana contain rare or surprising species that do not seem to be permanent residents of the groves.

### a) Draw River

This is an interesting forest in good condition and with a high genetic heat index (Table 3.1). Its conservation is highly desirable. Many previous observers have suggested that it should be integrated as a part of Ankasa/Nini Suhien, with which it is contiguous, and this seems a very reasonable suggestion. The elephant population actually migrates between Ankasa and Draw River (see for example Parren and de Graaf 1995 on movements of elephants in the area and proposed establishment of corridors).

### b) Boi-Tano

This is another forest in relatively good condition with a high botanical genetic heat index which can act as a stepping stone between Nini Suhien and Bia. Despite the very poor collecting during the RAP survey, there is no doubt that the forest has a full butterfly fauna. Encroachment does not seem a problem, but some restraint on hunting is needed. The central access road, an old logging road, is excessively broad now that there is no heavy traffic except seasonal cocoa trucks. The roads impede many arthropods from crossing from one part of the forest to the other and allow the growth of *Chromolaena odorata*; a nearly closed canopy above the road (obtained by keeping access roads in the reserves really narrow) would help eliminate this noxious weed.

### c) Krokosua Hills

Much of Krokosua is in a rather poor state. The botanical genetic heat index is low. Encroachment is probably quite widespread, not only in the area visited during the RAP mission. Hunting pressures are high. However, on the central ridge there is still some good forest and the butterfly fauna (three very rare and unusual species and a full forest floor fauna) indicates that it is much richer than in the damaged forest along the edges of the ridge. Possibly the ridge could be designated as a specially protected core area as a trade-off for continued human activity in the poorer lowland forests.

## REFERENCES

Dall'asta, U., J. Hecq and T.B. Larsen. 1994. L'emploi de papillons de jour (Insectes: Rhopalocera & Grypocera) comme 'espèces monitrices' et 'espèces indicatrices' dans le projet de rehabilitation des fôrets dans l'Est de la Côte d'Ivoire. Rapport Lepidoptera, No. 1. Musée Royal de l'Afrique Centrale (conference paper). 47 pp.

Emmel, T. and T.B. Larsen. 1997. Butterfly diversity in Ghana, West Africa. Tropical Lepidoptera, 8 (supplement 3):1-13.

Hawthorne, W.D. and M. Abu-Juam. 1995. Forest Protection in Ghana. IUCN/ODA/Forest Department Republic of Ghana, Gland, Switzerland, and Cambridge, UK, xvii + 203 pp.

Larsen, T.B. 2005. Butterflies of West Africa. Apollo Books, Svendborg, Denmark.

Larsen, T.B. 2001. The butterflies of Ankasa/Nini-Suhien and Bia protected area systems in western Ghana. Protected Areas Development Programme. UGL/Ghana Wildlife Department. 62pp.

Lawton, J.H., D.E. Bignell, B. Bolton, G.F. Bloemers, P. Eggleton, M. Hodda, R.D. Holt, T.B. Larsen, N.A. Mawdsley, N.E. Stork, D.S. Srivastava and A.D. Watt. 1998. Biodiversity inventories, indicator taxa and effects of habitat modification in tropical forest. Nature, 391:72-76.

Parren, M.P.E. and N.R. de Graaf. 1995. The quest for natural forest management in Ghana, Côte d'Ivoire and Liberia. The Tropenbos Foundation, Wageningen, 199 pp.

A biological assessment of the terrestrial ecosystems of the Draw River, Boi-Tano, Tano Nimiri and Krokosua Hills forest reserves, southwestern Ghana

39

# Chapter 4

## Rapid Assessment of Insects (excl. Butterflies) of Draw River, Boi-Tano and Krokosua Hills

*Frederick Ansah*

### SUMMARY

A survey of the insect fauna (excluding Lepidoptera) was undertaken at the Draw River, Boi-Tano and Krokosua Hills forest reserves in the Western Region of Ghana. Three sites were sampled over a period of 20 days. The methods of collection were principally the use of aerial sweep nets, pit-fall trapping and opportunistic observation of species. A total of 589 insect specimens belonging to 10 orders were collected. There were 56 families of insects collected or encountered, with Coleoptera being the dominant order with 13 families, followed by Heteroptera and Diptera with 11 families each, and Hymenoptera with nine families. Other orders in decreasing numerical order were Orthoptera (3), Odonata (3), Dictyoptera (3), Dermaptera (1), Isoptera (1), Neuroptera (1). The number of families of insects recorded at the three sites were: Boi-Tano (38), Draw River (31) and Krokosua Hills (34) with 93 identified species being found at the three sites (Appendix 4). Species diversity was highest at Boi-Tano with 59 species, followed by Krokosua Hills with 54 species and Draw River with 47 species. The survey indicated the presence of a wide selection of insect families, which should be systematically collected for further research and study in order to build a good taxonomic database for the West African sub-region.

### INTRODUCTION

In tropical conservation, decision-making regarding the demarcation of protected areas has been largely based on studies of "charismatic" vertebrates (e.g. elephants, lions, etc.). However conservation is not simply about saving mammals, nor do arguments from science, ethics and ecology provide the only rationale in support of it, even though the tourist industry in Africa is highly dependent on large mammals as a major attraction (Kingdon 1997). The large store of knowledge provided by specialists in mammal ecology has also helped in perpetuating the trend. The state of knowledge on invertebrates is scanty or largely non-existent in many tropical areas, with available information often being out of date or of poor quality, and the results also often available only in journals, museums, and universities, or in the minds of experts (Bakarr et al. 2001). A great many additional invertebrate species await discovery, identification and naming, thereby increasing the difficulties associated with tropical invertebrate studies. Taxonomic difficulties make it impossible to process or identify invertebrate species under field conditions, thus contributing to the inability to monitor the abundance, presence or absence of individual species. Such deficiencies lead to tragic results of endangerment or total extinction of some species before they are ever discovered, identified and named.

Insects, due to their short life span, are good indicators of environmental change (e.g. humidity, temperature, heavy metal contamination of soil and water, degradation and other activities of man). For example katydids (Tettigonidae) have been known to exhibit a remarkable potential as environmental indicators (Samways 1997). Many species of invertebrates are essential in the food web, serving as primary food sources for other animals, and playing major roles in nutrient recycling through leaf litter and wood degradation, carrion and dung disposal, and soil turnover. They also play major roles in plant pollination, and maintenance of plant community composition and structure through phytophagy (Gullan and Cranston 1994).

This report presents the results of a Rapid Assessment Program (RAP) survey of three forest reserves in Southwestern Ghana (Moist and Moist-Semi Deciduous Forests).

## Study sites

The study areas were the Draw River, and Boi-Tano forest reserves situated within the Wet Evergreen Forest (WEF) and the Krokosua Hills forest reserve within the Northwestern subtype of Moist Semi-Deciduous Forest. The Wet Evergreen Forest comprises large trees with vertically compressed canopies rarely exceeding 40 m. The mean annual rainfall is 1,750 mm to 2,000 mm. Boi-Tano Forest Reserve is a logging area while Draw River Forest Reserve has cocoa plantations and farmlands. The Moist Semi-Deciduous Forest (MS) comprises trees that are taller than those of any other forest type. The heights often exceed 50 m and sometimes 60 m. The upper canopies consist of deciduous and evergreen species in varying proportions. The mean annual rainfall is 1,500mm (Hall and Swaine 1976, 1981). Krokosua Hills Forest Reserve also has cocoa plantations and farmlands.

## Methodology

### Aerial Netting

Butterfly aerial nets were used to opportunistically collect flying insects, which were then transferred into a killing bottle lightly charged with ethyl acetate vapour. The insects were then enveloped for later sorting and pinning.

### Sweep Netting

Sweep nets were swung vigorously against shrubs, twigs, and thickets to dislodge specimens. Using standardized sweeping, there were five swings in a series before emptying the net into a container charged with ethyl acetate vapour. The specimens were separated from the broken twigs and leaves in a sorting tray for post treatment pinning or storage in 70% ethyl alcohol.

### Pitfall trapping

Pitfall trapping involved the sinking of medium sized plastic pails to the level of the topsoil in the forest floor. The traps contained 150 cm³ of water and a few drops of 40% formaldehyde solution. Apart from wetting the specimens, the solution also killed and preserved them as the traps were not inspected daily. Specimens collected were either pinned or stored in 70% alcohol.

### Direct Opportunistic Observation

There was visual observation of terrestrial, flying and surface-water dwelling insects. Collected specimens were pinned or stored in 70% alcohol.

The specimens were identified using voucher specimens in the Entomology Museum, Zoology Department, University of Ghana, Legon.

## Results

Collecting effort on each day was 3.5 hours of trapping and walking in addition to 1.5 hours for sorting, pinning and preservation in alcohol. Despite occasional wet weather, which disrupted sampling, especially at Krokosua, the three sites sampled yielded 589 specimens belonging to 10 orders. There were 57 families of insects collected or encountered, with Coleoptera being the dominant order with 13 families, followed by Heteroptera and Diptera with 11 families each, and Hymenoptera with nine families. Other orders in decreasing numerical order were Orthoptera (3), Odonata (3), Dictyoptera (3), Dermaptera (1), Isoptera (1), Neuroptera (1). The number of families of insects recorded at the three sites were: Boi-Tano (37), Draw River and Krokosua Hills (35 each) (Appendix 3). Of the specimens collected, we identified 93 identified species found at the three sites (Appendix 4). Among the identified species, diversity was highest at Boi-Tano with 60 species, followed by Krokosua Hills with 54 species and Draw River with 49 species.

### Aerial netting

The main species collected were from the orders Coleoptera, Diptera, Heteroptera, Hymenoptera, Odonata and Orthoptera. Butterflies were encountered in large numbers, but were not a target group and hence were not collected (Table 4.1).

**Table 4.1** Numbers of insects collected by aerial netting from the three sites

| Taxon | Draw River | Boi-Tano | Krokosua Hills | TOTAL |
|---|---|---|---|---|
| Coleoptera | 13 | 8 | 15 | 36 |
| Diptera | 16 | 16 | 18 | 50 |
| Heteroptera | 20 | 9 | 15 | 44 |
| Hymenoptera | 2 | 8 | 10 | 20 |
| Odonata | 3 | 4 | 31 | 38 |
| Orthoptera | 5 | 11 | 8 | 24 |

## Sweep Netting

A large number of insects were collected using this method, with predominant orders being Heteroptera and Hymenoptera (Formicidae and Apidae). Fewer numbers of Dictyoptera, Diptera, and Neuroptera were also collected (Table 4.2). Other arthropods of ecological importance found in the catch were spiders (Aranea).

## Pitfall trapping

Ground beetles or Coleoptera (*Analchalcus cupreus* and *Cyphonistes rufocasteneus*) and ants, Hymenoptera (*Palthothyreus* sp.), were the main insects collected in pitfall traps (Table 4.3). Other arthropods of ecological significance caught using this method were reduvids (Reduviidae, assassin bugs) and ground-dwelling spiders (Aranea). Skinks (Reptilia) and toads (Amphibia) were also occasionally found in the traps.

## Direct Opportunistic Observation

Tipulidae were found in the buttresses of trees. Simuliidae, Gyrinidae, Dysticidae and other aquatic insects were also observed.

## DISCUSSION

The insects collected showed a fair distribution of species in all the sites in spite of the farming activities in both Draw River and Krokosua Hills forest reserves (Table 4.4). In the order Coleoptera, the families Cerambycidae, Dysticidae, Gyrinidae and Scarabeidae are represented in all three sites. In the order Diptera, the families Drosophilidae, Muscidae, Simulidae and Tabanidae are represented. The order Hemiptera has the families Corixidae, Gerridae, Hydrometridae, and Reduviidae represented in all three sites. The order Hymenoptera has the families Apidae, Formicidae, Ichneumonidae and Scoliidae represented in the three sites. The order Orthoptera has two families, Acrididae and Tettigonidae, in all the sites surveyed. Families Libellulidae (Odonata)

and Mantidae (Dictyoptera) were found in all three sites. Forficulidae (Dermaptera) was encountered only at Krokosua Hills. The only Termitidae (Isoptera) was collected at Draw River. The 53 species of insects recorded for Draw River and 48 species for Krokosua Hills could be explained by the presence of large tracts of cocoa plantations and food crop farms at the two sites that were sprayed with pesticide at various times during the sampling period. Farmlands and access roads bordering and entering the forests also had vast stretches of *Chromolaena odorata* ("Acheampong weed"), which prevents some species migration from the forest into the fringes. This was particularly so at Boi-Tano where logging tracks were common. The results above show that the most efficient trapping technique is aerial netting.

For a more complete assessment, collection of insects should be done using other methods such as (i) light trapping, (ii) knockdown technique using pesticides (though it is noted that this method has adverse ecological consequences), and (iii) malaise traps. However, time constraints and limited staff did not enable the use of these methods during the RAP survey.

## CONSERVATION RECOMMENDATIONS

Under the circumstances, our methods, though not very exhaustive, provided evidence of the diversity of insects in the study area. In the light of the above, it is recommended that further work be undertaken to obtain more complete information. Boi-Tano Forest Reserve is threatened by the logging of timber in the reserve. Farming activities must be vigorously discouraged in and around the reserves to prevent the increasing pesticides usage, which destroy non-target species and therefore the ecosystem in the long term. The survey has shown the presence of a wide selection of insect families, which should be systematically collected for further research and study to build a good taxonomic database for collaborative work in the sub-region.

**Table 4.2.** Numbers of insects collected by sweep netting from the three sites

| Taxon | Draw River | Boi-Tano | Krokosua Hills | TOTAL |
|---|---|---|---|---|
| Dictyoptera | 13 | 3 | 1 | 17 |
| Diptera | 8 | 3 | 4 | 15 |
| Heteroptera | 20 | 8 | 23 | 51 |
| Hymenoptera | 3 | 9 | 14 | 26 |
| Neuroptera | 2 | - | 1 | 3 |

**Table 4.3.** Numbers of insects collected by pitfall traps from the three sites

| Taxon | Draw River | Boi-Tano | Krokosua Hills | TOTAL |
|---|---|---|---|---|
| Coleoptera | 32 | 31 | 20 | 83 |
| Hymenoptera | 7 | 17 | 19 | 43 |

**Table 4.4.** Summary of numbers of insects, by order, collected at the three sites

| Taxon | Draw River | Boi-Tano | Krokosua Hills | TOTAL |
|---|---|---|---|---|
| Coleoptera | 59 | 49 | 55 | 163 |
| Dermaptera | - | - | 2 | 2 |
| Dictyoptera | 13 | 4 | 1 | 18 |
| Diptera | 24 | 19 | 22 | 65 |
| Heteroptera | 40 | 17 | 38 | 95 |
| Hymenoptera | 12 | 34 | 43 | 89 |
| Isoptera | - | 1 | - | 1 |
| Odonata | 3 | 4 | 31 | 38 |
| Orthoptera | 25 | 49 | 36 | 110 |
| Neuroptera | 2 | - | 1 | 3 |
| **TOTAL** | **178** | **177** | **229** | **584** |

# REFERENCES

Bakarr, M., B. Bailey, D. Byler, R. Ham, S. Olivieri and M. Omland. 2001. From the forest to the sea: biodiversity connections from Guinea to Togo. Conservation Priority-Setting Workshop, December 1999. Conservation International, Washington, DC, 78 pp.

Gullan, P.J. and P.S. Cranston. 1994. The Insects: An Outline of Entomology, Chapman and Hall, London. xiv + 471 pp.

Hall, J.B. and M.D. Swaine. 1976. Classification and ecology of closed-canopy forest in Ghana. Journal of Ecology, 64:913-915.

Hall, J.B. and M.D. Swaine. 1981. Distribution and ecology of vascular plants in a tropical rain forest - Forest vegetation in Ghana. Dr W. Junk Publishers, The Hague. xv + 382 pp.

Kingdon, J. 1997. The Kingdon field guide to African mammals. Academic Press, San Diego, London. xviii + 464 pp.

Samways, M.J. 1997. Conservation Biology of Orthoptera. Pp. 481-496. *In:* Gangvere, S.K. et al. (eds.). The Bionomics of grasshoppers, katydids and their kin. CAB International.

# Chapter 5

## Herpetological assessment of Draw River, Boi-Tano, and Krokosua Hills

*Raffael Ernst, Alex Cudjoe Agyei and Mark-Oliver Rödel*

## SUMMARY

We investigated the herpetofauna of portions of three forest reserves, designated as Globally Significant Biodiversity Areas (GSBAs) in the Western Region of Ghana. We recorded a total of 39 amphibian species, among them the first country record for the genus *Acanthixalus*, as well as a new species of the genus *Arthroleptis*. The species *Acanthixalus sonjae* was previously only known from Côte d'Ivoire. *Phrynobatrachus annulatus* is reported for the first time from Ghana. *Phrynobatrachus ghanensis* is reported for the first time outside of the Bobiri and Kakum forest reserves. The 18 reptile species recorded include three species of global conservation concern (*Chamaeleo gracilis*, *Osteolaemus tetraspis* and *Kinixys erosa*). Reptiles were underrepresented due to lower detectability in the field. This holds especially true for snakes. Most of the recorded species were either endemic to West Africa or even to the Upper Guinean forest block.

The relatively high diversity and/or unique species composition with respect to regional endemicity documented during this RAP survey, most prominent in Draw River, clearly demonstrates that these areas still have a high potential for nature conservation. However, we also documented several invasive species (*Hoplobatrachus occipitalis*, *Phrynobatrachus accraensis*, *Agama agama*), normally not occurring in forest areas, which clearly indicates significant alteration of the original forest habitats by means of unsustainable forest use (encroachment / illegal farming, illegal timber extraction, etc.).

## INTRODUCTION

Despite the fact that West Africa has been the target of herpetological investigations for more than 100 years, our present knowledge is still scanty (Hughes 1988 provides an overview on the history of herpetological investigations in Ghana). Since the work by A. Schiøtz in Ghana, conducted in the 1960s (Schiøtz 1964a, b, 1967), only a very few papers have been published on Ghanaian herpetology. A few forest surveys have revealed between 10 and 20 amphibian species per site (for a recent summary see Rödel and Agyei 2003). The forests of western Ghana have never been the target of a thorough herpetological survey. However, recent investigations in Côte d'Ivoire and Guinea have revealed anuran communities comprising more than 30 species, even in savannah habitats. Well studied forest communities comprise between 40 and 60 species (Ernst and Rödel 2002, Rödel 2000, 2003; Rödel and Spieler 2000; Rödel and Branch 2002; Rödel and Ernst 2003, 2004; Rödel and Bangoura 2004; Rödel et al. 2004). It is very likely that the Ghanaian forests are not less diverse than Ivorian or Guinean ones, but are simply less well explored.

For this reason, A. Schiøtz and M.-O. Rödel defined the western Ghanaian forests as an area with an exceptionally high priority for rapid assessment during the Conservation Priority Setting Workshop in Ghana (Bakarr et al. 2001a). As an outcome of that workshop, Conservation International conducted a Rapid Assessment Program (RAP) survey in three selected forests of south-western Ghana. The herpetological results are presented herein.

## METHODS

### Sampling methods and effort

Specimens were mainly located opportunistically, during visual surveys (VES) of all habitats by up to three members. Surveys were undertaken during the day and during the evening (until 2200h). Search techniques included visual scanning of terrain (using flashlight by night) and refuge examination (e.g. lifting rocks and logs, peeling away bark, scraping through leaf litter). Additional specimens or sightings were provided by other members of the biological inventory team, as well as local inhabitants. Amphibians were also investigated by acoustic monitoring of all available habitat types (AES; Heyer et al. 1993).

To supplement opportunistic collecting, habitats were also sampled using arrays of pitfall traps placed along drift fences. Trap lines (total of three per survey site) were set in different microhabitat types (Table 5.1). Drift fences consisted of lengths of plastic shade cloth or black plastic sheeting 0.5 m high stapled vertically onto wooden stakes. An apron left at the base was covered with soil and leaf litter to ensure specimens intercepted during their normal movements moved along the fence towards the traps. Pitfall traps comprised plastic water buckets (275 mm deep, 285 mm top internal diameter, 220 mm bottom internal diameter) sunk with their rims flush with ground level and positioned so that the drift fences ran across the mouth of each trap. One pitfall trap was set at each end of a drift fence with other traps spaced in between at regular intervals (total of six buckets per line). Holes in the base of the buckets allowed drainage. Traps were checked every morning and during the day if a survey team was working in the region. Specimens not retained as voucher specimens were released in the vicinity of capture. Drift fence lengths, orientation and trap arrays were tailored to local conditions. Each trap array comprised a single straight line fence (15 m) with six buckets. Trap lines were set for variable periods (Table 5.1). A trap-day is defined as one trap in use for a 24-hour period.

Some voucher specimens were collected. Amphibian vouchers were anesthetized and killed using toothache pain relief gel containing 20% Benzocaine. Reptiles were killed using Halothane. All vouchers were preserved in 70% ethanol. Vouchers will be deposited in various museum collections and the research collection of M.-O. Rödel (amphibians: Staatliches Museum für Naturkunde Stuttgart, Germany and MOR; reptiles: Port Elizabeth Museum, South Africa).

## RESULTS

### Trapping and Survey Results

Since only two of the recorded reptile species (*Hemidactylus fasciatus* and *Hemidactylus muriceus*) potentially vocalize, all records of reptiles were sightings and/or captures. With amphibians, however, AES was an important tool, especially for most arboreal species that vocalize frequently but are rarely encountered otherwise (e.g. *Leptopelis* spp.). However, 39 amphibian species belonging to seven different families were recorded employing VES, whereas AES only yielded 24 species, belonging to three different families. Bucket traps with drift fences captured seven species belonging to five different families. Among those were two species from two different families not recorded otherwise (*Geotrypetes seraphini* and *Silurana tropicalis*; see Figure 5.1). Drift fences also yielded records of four different species of lizards, one of which (*Mochlus fernandi*) was not recorded otherwise. One frog family (Astylosternidae) was recorded by dip-netting only.

Trapping success for bucket drift fence arrays was variable, depending on particular locations. A total of 37 specimens were collected in traps, comprising 31 amphibians and

**Table 5.1:** Details of drift fence arrays at Draw River (DR), Boi-Tano (BT), and Krokosua Hills (KH)

| SITE | LENGTH (m) | PITFALL TRAPS | DAYS | HABITAT |
|------|-----------|---------------|------|---------|
| DRP1 | 15 | 6 | 5 | Undulating terrain, closed canopy forest near stream |
| DRP2 | 15 | 6 | 5 | Dry uphill area, closed canopy forest app. 150 m away from stream |
| DRP3 | 15 | 6 | 4 | Poorly drained low-lying open area with dense undergrowth |
| BTP1 | 15 | 6 | 4 | Vicinity of small stagnant forest creek in swampy area along reserve boundary |
| BTP2 | 15 | 6 | 4 | Dry closed forest area without any apparent bodies of water |
| BTP3 | 15 | 6 | 4 | Dry closed forest area without any apparent bodies of water |
| KHP1 | 15 | 6 | 7 | Next to precut line at first ascent of a larger hill (top app. 500 m a.s.l.) in close proximity to illegal plantation |
| KHP2 | 15 | 6 | 7 | Next to precut line at first ascent of a larger hill (top app. 500 m a.s.l.) in close proximity to illegal plantation |
| KHP3 | 15 | 6 | 5 | Low-lying wet forest, vicinity of dried out creek |

6 lizards. Amphibian and reptile captures are summarized in Tables 5.2 and 5.3, respectively.

*Sampling efficiency*
With the kind of sampling design employed, only qualitative and semi-quantitative data can be obtained. For exact

**Table 5.2:** Amphibians caught in trap arrays at Draw River, Boi-Tano, and Krokosua Hills

| Species | Number |
|---|---|
| *Bufo maculatus* | 16 |
| *Arthroleptis* sp. 1 & 2 | 10 |
| *Phrynobatrachus alleni* | 2 |
| *Phrynobatrachus annulatus* | 1 |
| *Silurana tropicalis* | 1 |
| *Geotrypetes seraphini* | 1 |
| **Total** | **31** |

**Table 5.3:** Reptiles caught in trap arrays at Draw River, Boi-Tano, and Krokosua Hills

| Species | Number |
|---|---|
| *Mabuya affinis* | 2 |
| *Lygodactylus conraui* | 2 |
| *Panaspis togoensis* | 1 |
| *Mochlus fernandi* | 1 |
| **Total** | **6** |

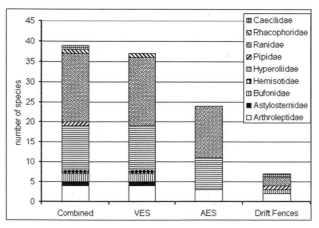

**Figure 5.1:** Number of species recorded using a particular sampling method. Combined = VES + AES + Drift Fences with Bucket Traps.

quantitative data, mark-recapture experiments along standardized transects or on definite plots would have been necessary. Since this survey was restricted to approximately one week at each site, time was not sufficient to employ these methods. Species accumulation curves show number of new species added each day (Figure 5.2). A continued increase of the curve's slope indicates that additional amphibian and reptile species remain to be recorded within the region. To evaluate the sampling effort we measured the time spent searching at each locality (man hours - m/h).

Assuming that sampling effort was the same for each habitat, we calculated the approximate total number of amphibian and reptile species occurring at the site (Figures 5.3 and 5.4). Because we only had qualitative data available, we used the Jack-knife 1 estimator, based on presence/absence data (program: BiodivPro http://www.sams.ac.uk/dml/projects/benthic/bdpro/index.htm). No estimator values were calculated for reptiles at Boi-Tano, due to data deficiency. Recorded species richness was generally lower than the calculated estimates for any given site (Figures 5.3 and 5.4). Steep slopes, especially in the Draw River reptile estimator's curve, indicate that not all species present in the region were recorded, and that the calculated number of only 13 reptile species present is too low.

*Status of the amphibian faunas of Draw River (DR), Boi-Tano (BT), and Krokosua Hills (KH) forest reserves*
We recorded 28 amphibian species in Draw River, 22 species in Boi-Tano, and 23 species in Krokosua Hills, respectively (see Appendix 6). With a total of 39 species recorded, the species richness of the area under investigation can be considered moderate to high. This is especially true when compared to the most diverse West African region so far investigated, namely Taï National Park in southwestern Côte d'Ivoire. Although the recorded species richness in the Ghanaian forests surveyed does not reach the level of Taï NP, there is a high resemblance between the species composition of the areas surveyed and other West African sites. In

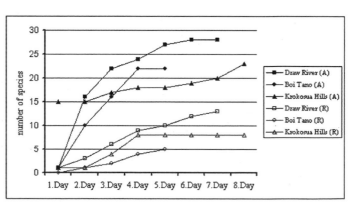

**Figure 5.2:** Species accumulation curves for Draw River, Boi-Tano, and Krokosua Hills forest reserves (A=amphibians; R=reptiles)

a species similarity analysis with other known West African amphibian assemblages (M.-O. Rödel and M. Wegmann unpubl. data), Krokosua Hills groups with Bobiri forest reserve in Ghana. These both are placed together with sites that comprise savannah and drier forest environments. Draw River and Boi-Tano group together with each other in a cluster that comprises West African forest sites with moderate to intensive level of degradation and moderate species richness (Mont Péko NP, Haute Dodo forest reserve, Cavally forest reserve, Rödel and Branch 2002, Rödel and Ernst 2003). In all three areas the vast majority of the species were either Upper Guinean rain forest endemics or at least restricted to West Africa (defined as the region from Senegal to Eastern Nigeria). Only a very few species, three in Draw River and Krokosua Hills and two in Boi-Tano, have wider distributions in Sub-Saharan Africa.

A number of notable species were recorded during our survey. In Draw River we found *Geotrypetes seraphini*, a rarely recorded caecilian of uncertain conservation status, as well as an endemic Ghanaian frog, *Phrynobatrachus ghanensis*, which was previously known only from Kakum and Bobiri forest reserves. *Phrynobatrachus annulatus* was recorded from Draw River and Boi-Tano and thus extended the known range of this species, so far believed to be a western Upper Guinea endemic. At Krokosua Hills, the hyperoliid genus *Acanthixalus* was recorded for the first time in Ghana. The recorded specimens were genetically identical (only 3 substitutions in the 16S rRNA gene, M.-O. Rödel & J. Kosuch unpubl. data) compared to *Acanthixalus sonjae* specimens from western Côte d'Ivoire. This species was very recently described from Taï National Park (Rödel et al. 2003). An *Astylosternus* tadpole from Draw River may represent the West African *Astylosternus occidentalis* or a new species (assumption based on pictures from adjacent Ankasa NP suggesting an apparently new species of this genus, pictures kindly provided by M. Gill). Additionally an *Arthroleptis* species

was found at the slopes of a hill at Krokosua Hills, which is a new species to science. The locality records and voice recordings of several difficult frog groups collected during the surveys will help to clarify taxonomic problems in the *Hyperolius fusciventris* species group (in which a number of subspecies are probably best regarded as full species). DNA-samples will be of similar importance in the *Ptychadena aequiplicata* complex, as well as the new *Arthroleptis* record.

In addition to the discovery of forest specialists, a number of typical farmbush species (e.g. *Afrixalus dorsalis*, *Hyperolius concolor*) or even savannah species (*Hoplobatrachus occipitalis*, *Phrynobatrachus accraensis*, *Bufo regularis*) were recorded. This indicates that all three forest reserves have already been invaded by species that are not normally present.

### Status of the reptile faunas of of Draw River (DR), Boi-Tano (BT), and Krokosua Hills (KH) forest reserves

Although only ten species of lizards were recorded across all sites surveyed (DR: 8, BT: 3, KH: 6), the species richness of the region lies within the average range recorded in similar West African rain forest sites. Generally, forest lizard faunas in West Africa are known to be of very low diversity (Rödel et al. 1997, Branch and Rödel 2003). Böhme (1994) recorded only 11 species from Ziama classified forest, and an equally impoverished fauna is known from Mount Nimba (15 species, Angel et al. 1954a, b, Böhme et al. 2000, Ineich 2002). Whilst Lawson (1993) noted affinities between the snake faunas of Korup, Cameroon and the Upper Guinea Forests, few such relationships occur between the lizard faunas. Most lizards have relatively wide distributions within the region (e.g. *Hemidactylus fasciatus*, *H. muriceus*, *Agama agama*, *Chamaeleo gracilis*, *Mabuya affinis*, etc.). The only lizards endemic to the Upper Guinea forests are the water skink *Cophoscincopus durus* and the western race of the forest skink, *Mabuya polytropis paucisquamis* (which probably deserves specific recognition, see Branch and Rödel 2003). The two chameleons (both

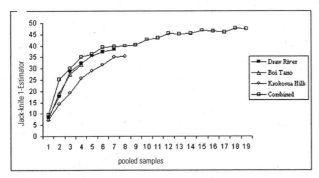

**Figure 5.3:** Jack-knife Estimator for species richness in amphibians. Combined = pooled data from Draw River, Boi-Tano, and Krokosua Hills. Estimated species richness Draw River: 39, recorded: 28 (71.8 %); Boi-Tano: 32, recorded: 22 (68.8 %); Krokosua Hills: 35, recorded: 23 (65.7 %); Combined: 48, recorded: 39 (81.3 %)

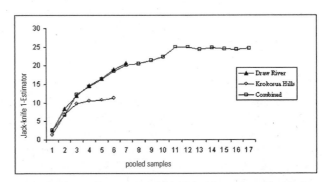

**Figure 5.4:** Jack-knife Estimator for species richness in reptiles. Combined = pooled data from Draw River, Boi-Tano, and Krokosua Hills. Estimator for Boi-Tano is not depicted due to data deficiency. Estimated species richness Draw River: 21, recorded: 13 (61.9 %); Krokosua Hills: 11, recorded: 8 (72.7 %); Combined: 25, recorded: 18 (72.0 %)

*Chamaeleo gracilis*) were caught by local inhabitants of the village of Mmem in Krokosua Hills. They reported the use of dried specimens for medicinal and ceremonial purposes. Although we did not record any monitor lizards (*Varanus* sp.) during our survey, local hunters at Krokosua Hills reported that they are hunted for meat.

With only six species recorded, species richness in snakes was even lower than species richness in lizards. However, this is most certainly due to undersampling. A higher sampling intensity over longer periods of time would most likely yield additional species. Based on known snake faunas from other West African regions (Böhme 1999, Rödel and Mahsberg 2000, Ineich 2002) a fauna of about 40 snake species should occur in these forests. Endemicity in snakes is generally relatively low. Nonetheless, four of six species recorded in this survey are endemic to the Upper Guinea forests. These include the poorly-known *Lycophidion nigromaculatum, Dendroaspis viridis,* and *Thrasops* sp.

One crocodilian species (*Osteolaemus tetraspis*), recorded from Draw River and Boi-Tano/Tano Nimiri, and one tortoise species (*Kinixys erosa*), recorded from Draw River, are currently considered threatened. They are both included on the 2004 IUCN Red List and listed as Vulnerable and Data Deficient, respectively (IUCN 2004). *Chamaeleo gracilis* is included in CITES Appendix II and its international trade is monitored or regulated. Sightings of chelonians and crocodilians were rare. *Osteolaemus tetraspis* was only seen once at Draw River and Tano Nimiri (for the latter: Abdulai Barrie pers. com.). The only records of *Kinixys erosa* were an adult male and a juvenile, both kept by local villagers at Draw River and intended to be used as food. Local villagers also reported hunting both *Osteolaemus tetraspis* and *Kinixys erosa* for food. Lawson (2000) has documented the extent of exploitation of the forest tortoises *Kinixys erosa* and *K. homeana* in Cameroon, and the dwarf crocodile *Osteolaemus tetraspis* is similarly threatened by overexploitation throughout West Africa (Akani et al. 1998).

## DISCUSSION AND CONSERVATION RECOMMENDATIONS

In contrast to large mammals and/or primates, habitat size per se seems to be less important than habitat diversity in conserving amphibian diversity (Ernst and Rödel unpubl. data). This may be one of the reasons why both diversity and species composition in the areas under investigation indicate high potential for preserving amphibian and reptile species typical for the Upper Guinean forest region. The species composition in the amphibian assemblages partially resembles that of other West African sites, but also shows unique characters in species composition, typical for the eastern part of the Upper Guinea forest zone. However, the presence of typical farmbush and even savannah species is a clear hint that the respective habitats are already seriously damaged. Habitat loss through illegal plantations and encroachment is a major concern in all of the areas visited. It is most dramatic at Krokosua Hills, where virtually all low-lying forests were

disturbed due to well established illegal plantations (mainly subsistence farming and cocoa plantations), with only forest on steep slopes remaining relatively intact. Continued logging activities also lead to high levels of disturbance and severe alteration in microclimate and water balance. Due to their physiology and biphasic lifecycle, amphibians are especially susceptible to these alterations and thus prone to local extinctions. Although habitat size may not be of utmost importance for the protection of amphibians, habitat diversity, which is important, is likely to increase with habitat size. A minimum habitat size is also essential to maintain ecosystem functions (e.g. climatic conditions) that determine and thereby directly affect amphibians. Our observations clearly show that some invasive species are already well established within the forests reserves. It is not unlikely that invasives will compete and eventually displace true forest species.

Highest priority should thus be given to halting or strictly controlling ongoing encroachment and illegal farming activities, as well as logging. One legislative issue requiring particular attention involves the common practice of "legalizing" plantations even within clearly demarcated reserves by planting the cash crop cocoa. In our understanding, these illegal cocoa plantations cannot be replaced or destroyed once they are established. This, however, will pave the way for continuous destruction of the remaining forest habitats. Improvement of infrastructure (especially road construction) around the reserves is another serious concern. It is now well established that access to previously inaccessible forest by logging roads leads to numerous impacts, including a massive increase in the illegal and unsustainable harvesting of forest resources, especially bushmeat (see Bakarr et al. 2001b and references therein). It also allows easy access to forest sites and therefore facilitates the aforementioned establishment of small illegal cocoa plantations. Both of these major impacts are already very evident in the forest reserves. Although poaching may not be a direct threat to most amphibians, it strongly affects a number of reptiles, among them species included in the Red List and regulated under CITES (*Osteolaemus tetraspis, Kinixys erosa, Varanus* spp., and potentially *Python* spp.). Many species are regularly hunted for food or for medicinal and/or ceremonial purposes (e.g. *Chamaeleo gracilis*). Hunting pressure on reptiles is likely to increase due to severe exploitation and recent lack of more traditional game such as large- and medium-sized mammals, primates and larger birds. Active protection of these taxa will therefore lead to decreasing hunting pressures on reptiles and thus serve as a direct protection measure.

## REFERENCES

Akani, G.C., L. Luiselli, F.M. Angelici and E. Politano. 1998. Bushmen and herpetofauna: notes on amphibians and reptiles traded in bush-meat markets of local people in the Niger Delta (Port Harcourt, Rivers State, Nigeria). Anthropozool., 27: 21-26.

Angel, F., J. Guibe and M. Lamotte. 1954a. La réserve naturelle intégrale du mont Nimba. Fasicule II. XXXI. Lézards. Mem. Inst. fond. Afr. noire, sér. A, 40: 371-379.

Angel, F., J. Guibe, M. Lamotte and R. Roy. 1954b. La réserve naturelle intégrale du mont Nimba. Fasicule II. XXXII. Serpentes. Bull. Inst. fond. Afr. noire, sér. A, 40: 381-402.

Bakarr, M., B. Bailey, D. Byler, R. Ham, S. Olivieri and M. Omland. eds. 2001a. From the forest to the sea: Biodiversity connections from Guinea to Togo. Conservation Priority-Setting Workshop, December 1999, Washington DC. Conservation International, 78 pp.

Bakarr, M., G.A.B. DeFonseca, R. Mittermeier, A.B. Rylands and K.W. Painemilla (eds.) 2001b. Hunting and bushmeat utilization in the African rain forest. Perspectives toward a blueprint for conservation action. Adv. App. Biodiv. Sci., 2: 1-170.

Böhme, W. 1994. Frösche und Skinke aus dem Regenwaldgebiet Südost-Guineas, Westafrika. II. Ranidae, Hyperoliidae, Scincidae; faunistisch-ökologische Bewertung.- herpetofauna, 16 (93): 6–16.

Böhme, W. 1999. Diversity of a snake community in a Guinean rain forest (Reptilia, Serpentes). pp. 69-78. *In:* Rheinwald, G. (ed.). Isolated Vertebrate Communities in the Tropics. Proceedings of the 4th International Symposium, Bonner zoologische Monographien, 46.

Böhme, W., A. Schmitz and T. Ziegler. 2000. A review of the West African skink genus *Cophoscincopus* Mertens (Reptilia: Scincidae: Lygosominae): resurrection of *C. simulans* (Valliant, 1884) and description of a new species. Rev. Suisse Zool., 107: 777–791.

Branch, W.R. and M.-O. Rödel. 2003. Herpetological survey of the Haute Dodo and Cavally forests, western Ivory Coast, Part II: Trapping results and reptiles. Salamandra, 39: 21-38.

Ernst, R. and M.O. Rödel. 2002. A new *Atheris* species (Serpentes: Viperidae) from Tai National Park, Ivory Coast. Herpetological Journal, 12: 55-61.

Heyer, W.R., M.A. Donnelly, R.W. McDiarmid, L.-A.C. Hayek and M.S. Foster. 1993. Measuring and monitoring biological diversity, standard methods for amphibians. Washington DC. Smithsonian Institution Press, 364 pp.

Hughes, B. 1988. Herpetology in Ghana (West Africa). Brit. Herp. Soc. Bull., 25: 29-38.

IUCN 2004. 2004 IUCN Red List of Threatened Species. www.redlist.org

Ineich, I. 2002. Diversité spécifique des reptiles du Mont Nimba. unpublished manuscript.

Lawson, D.P. 1993. The reptiles and amphibians of the Korup National Park project, Cameroon. Herpetol. Nat. Hist., 1: 27-90.

Lawson, D.P. 2000. Local harvest of hingeback tortoises, *Kinixys erosa* and *K. homeana*, in southwestern Cameroon. Chelon. Conserv. Biol., 3: 722-729.

Rödel, M.-O. 2000. Herpetofauna of West Africa, Vol. I: Amphibians of the West African savanna. Edition Chimaira, Frankfurt/M., 335 pp.

Rödel, M.-O. 2003. The amphibians of Mont Sangbé National Park, Ivory Coast. Salamandra, 39: 91-110.

Rödel, M.-O. and A.C. Agyei. 2003. Amphibians of the Togo-Volta highlands, eastern Ghana. Salamandra, 39: 207-234.

Rödel, M.-O. and M.A. Bangoura. 2004. A conservation assessment of amphibians in the Forêt Classée du pic de Fon, Simandou Range, southeastern republic of Guinea, with the description of a new *Amnirana* species (Amphibia Anura Ranidae). Tropical Zoology 17: 201-232.

Rödel, M.-O. and W.R. Branch. 2002. Herpetological survey of the Haute Dodo and Cavally forests, western Ivory Coast, Part I: Amphibians. Salamandra, 38: 245-268.

Rödel, M.-O. and R. Ernst. 2003. The amphibians of Marahoué and Mont Péko National Parks, Ivory Coast. Herpetozoa, 16: 23-39.

Rödel, M.-O. and R. Ernst. 2004. Measuring and monitoring amphibian diversity in tropical forests. I. An evaluation of methods with recommendations for standardization. Ecotropica, Ulm; 10: 1-14.

Rödel, M.-O. and D. Mahsberg. 2000. Vorläufige Liste der Schlangen des Tai-Nationalparks / Elfenbeinküste und angrenzender Gebiete. Salamandra, 36: 25-38.

Rödel, M.-O. and M. Spieler. 2000. Trilingual keys to the savannah-anurans of the Comoé National Park, Ivory Coast. Stuttgarter Beiträge zur Naturkunde, Serie A, Nr. 620: 1-31.

Rödel, M.-O., M.A. Bangoura and W. Böhme. 2004. The amphibians of south-eastern Republic of Guinea (Amphibia: Gymnophiona, Anura). Herpetozoa 17 (3/4): 99-118.

Rödel, M.-O., K. Grabow, J. Hallermann and C. Böckheler. 1997. Die Echsen des Comoé-Nationalparks, Elfenbeinküste. Salamandra, 33: 225-240.

Rödel, M.-O., J. Kosuch, M. Veith and R. Ernst. 2003. First record of the genus *Acanthixalus* Laurent, 1944 from the Upper Guinean rain forest, West Africa, with the description of a new species. J. Herpetol., 37: 43-52.

Schiøtz, A. 1964a. A preliminary list of amphibians collected in Ghana. Vidensk. Meddr dansk naturh. Foren., 127: 1–17.

Schiøtz, A. 1964b. The voices of some West African amphibians. Vidensk. Meddr. dansk naturh. Foren., 127: 35–83.

Schiøtz, A. 1967. The treefrogs (Rhacophoridae) of West Africa. Spolia zool. Mus. haun., 25: 1–346.

A biological assessment of the terrestrial ecosystems of the Draw River, Boi-Tano, Tano Nimiri and Krokosua Hills forest reserves, southwestern Ghana

49

# Chapter 6

## Rapid assessment of the birds of Draw River, Boi-Tano and Krokosua Hills

*Hugo J. Rainey and Augustus Asamoah*

## SUMMARY

During 17 days of field work, 170 bird species were recorded: 126 in Draw River forest reserve, 109 in Boi-Tano forest reserve and 138 in Krokosua Hills forest reserve. Of these, five were of conservation concern (three in Draw River, two in Boi-Tano and four in Krokosua Hills). The total numbers of species now known from these sites are respectively: 142, 153 and 146 for Draw River, Boi-Tano and Krokosua Hills. Of the 15 restricted-range species that make up the Upper Guinea Forest Endemic Bird Area, four were found in Draw River, three in Boi-Tano and five in Krokosua Hills. A substantial component of the forest-restricted species in the country was found, as 96 of the 180 species of the Guinea-Congo Forests biome occurring in Ghana were recorded in Draw River, 82 in Boi-Tano and 97 in Krokosua Hills. The total numbers of forest biome species known from each site are 111, 114 and 102. This places all three sites in the top six forests in Ghana in terms of the numbers of forest biome birds recorded. Krokosua Hills now qualifies as an Important Bird Area, alongside Draw River and Boi-Tano. Our surveys have found the latter two forests to be more important than previously thought. Considering the conservation value of all three forests, it is recommended that further surveys be conducted in order to complete the species lists. Management and conservation recommendations are presented.

## INTRODUCTION

Birds have been proven to be useful as indicators of the biological diversity of a site. Their taxonomy and global geographical distribution are relatively well documented in comparison to other taxa (ICBP 1992), which facilitates their identification and permits rapid analysis of the results of an ornithological study. The conservation status of most species having been reasonably well assessed (BirdLife International 2000), the results and conclusions of such a study can be implemented productively. Birds are also among the most charismatic species, which can aid in the presentation of conservation recommendations to policy makers and stakeholders.

Previous studies of some of the remaining forests in West Africa have shown that they are of considerable importance for the survival of the birds of the Upper Guinea forests (e.g. Allport et al. 1989; Dutson and Branscombe 1990; Gartshore et al. 1995; Ntiamoa-Baidu et al. 2001; Demey and Rainey 2004, 2005). However, the avifaunas of the majority of the rapidly decreasing forests in West Africa remain inadequately known. In 1999 the forests of southwest Ghana were selected by experts as being of exceptionally high importance for the conservation of biodiversity (Bakarr et al. 2001). Accordingly, the following sites were chosen for study on the basis of their potential for conservation.

We carried out 17 days of field work at three sites: six days in Draw River forest reserve (FR) (22 – 28 October 2003), four days in Boi-Tano FR (29 October – 1 November) and seven days in Krokosua Hills FR (3 -9 November). All three sites had previously been the subject of bird surveys by the Ghana Wildlife Society as part of their Important Bird Area (IBA) program (Ntiamoa-Baidu et al. 2001). Both Draw River and Boi-Tano were already listed as

IBAs (Ntiamoa-Baidu et al. 2001) although some of this data was based on relatively old surveys from the 1980s (e.g. Dutson and Branscombe 1990). Accordingly we aimed to complement this previous work. Our efforts were focused on finding species of global conservation concern, Upper Guinea endemics and Guinea-Congo Forests biome species (species restricted to the forest zone in Ghana) (see Fishpool and Evans 2001 for further definitions). These categories can be used to define the importance of a site for birds (e.g. as an IBA). We also noted those species that can indicate forest disturbance and degradation.

For the purposes of standardization, we have followed the nomenclature, taxonomy and sequence of Borrow and Demey (2001). English names of species are given in Appendix 8 except for those not recorded from these forests (in this or previous surveys). These are given in the text.

## METHODS

The principal method used during this study consisted of observing birds by walking slowly along logging tracks and trails within the forests. Notes were taken on both visual observations and bird vocalizations. Tape-recordings of unknown vocalizations and those of rare species were made for later analysis and deposition in sound archives. Attempts were made to visit as many habitats as possible, particularly those that appeared likely to hold threatened or poorly known species. However, visits to more remote areas and some habitats (e.g. rivers) were limited by time and accessibility; it was therefore not possible to assess the abundance of some species found in these habitats. Similarly, time constraints at the second site (Boi-Tano/Tano Nimiri) forced us to spend almost all time in Boi-Tano; only one day was spent on the boundary separating Boi-Tano and Tano Nimiri. However, as these forests are contiguous, our results from Boi-Tano should be applicable to Tano Nimiri, at least in areas near to their common boundary. We have not, however, compared our results to previous work in Tano Nimiri.

Fieldwork was carried out from just before dawn (usually 05:30) until midday, and from 15:00 until sunset (around 18:00). Some further work was carried out at night to obtain records of owls and other species active at night. Typically this was carried out for a few hours after dusk. At Draw River and Krokosua Hills, we used view points from which the forest canopy could be observed to record many species that would often be difficult to see from the forest floor.

Mist-netting was carried out at all three sites. We used nine mist nets ranging in length from 10 to 18 m with a total combined length of 116 m. In addition, at Draw River we trapped birds using three nets set by the small mammal group for trapping bats at night, an additional 46 m. At Draw River we carried out 30.8 hundred meter net hours of trapping, at Boi-Tano/Tano Nimiri we ran 22.0 hundred meter net hours and at Krokosua Hills we ran 34.8 hundred meter net hours. Net sites at all three reserves were carefully selected taking into consideration the condition of the vegetation. Degraded areas as well as areas with clear signs of human presence such as active footpaths and hunters trails were avoided. We trapped for two to three days at each trapping site in each reserve. Nets were opened between 05:30 and 06:00 until 17:00 except on the second or third day of trapping when nets were taken down at about 14:00.

For each field day a list was compiled of all recorded species. Numbers of individuals or flocks were noted, as well as any evidence of breeding (e.g. the presence of juveniles) and basic information on the habitat in which the birds were observed. This enabled us to produce indices of abundance for each species based on the encounter rate (numbers of days on which a species was encountered and number of individuals and flocks involved). Comparisons can thus be made between the three sites and other sites in the region. The definitions of the abundance ratings are given in Appendix 8.

## RESULTS

We recorded a total of 170 species during our surveys in all three sites (see Appendix 8).

### Draw River Forest Reserve

During six days of fieldwork we recorded 126 species in this forest (see Appendix 8) amongst which were three species of global conservation concern: *Bleda eximia*, *Criniger olivaceus* and *Lamprotornis cupreocauda* (BirdLife International 2000; Table 6.1). These three species plus *Apalis sharpii* are all restricted-range species endemic to the Upper Guinea forests. We observed 96 Guinea-Congo forest biome species, 53% of the 180 forest biome species known from Ghana (Ntiamoa-Baidu et al. 2001). This is a large component of the forest-restricted species in the country. The total number of bird species now known from Draw River is 142 including 111 forest biome species (62%) (Ntiamoa-Baidu et al. 2001).

Mist-netting at Draw River was carried out on 24-25 October within mature secondary forest with little sign of recent disturbance: it was most recently logged in 1991 (Hawthorne and Abu-Juam 1995). Birds trapped at this site included *Bleda eximia*, and other species such as *Alcedo leucogaster*, *Phyllastrephus icterinus* and *Criniger barbatus* that are typical of mature forest. Eight species were trapped at a rate of 1.2 individuals per 100 meter net hours (see Appendix 9).

In addition, a number of species were observed at Draw River that are rare and poorly known in either Ghana or the Upper Guinea region. These included *Dryotriorchis spectabilis*, *Accipiter erythropus*, *A. melanoleucus*, *Otus icterorhynchus*, *Bubo poensis*, *Rhaphidura sabini*, *Neafrapus cassini*, *Prodotiscus insignis*, *Cinnyris johannae*, *C. batesi* and *Parmoptila rubrifrons*. Breeding activity of seven species was observed including nesting *Lamprotornis cupreocauda* (see Appendix 8).

## Boi-Tano/Tano Nimiri

Over four days we recorded 109 species in this forest (see Appendix 8) amongst which were two species of global conservation concern: *Bleda eximia* and *Illadopsis rufescens* (BirdLife International 2000; Table 6.1). These two species plus *Apalis sharpii* are all restricted-range species endemic to the Upper Guinea forests. We observed 82 Guinea-Congo forest biome species, 45% of the 180 forest biome species known from Ghana (Ntiamoa-Baidu et al. 2001). This is a relatively large component of the forest-restricted species in the country considering the time available. The total number of bird species now known from Boi-Tano is 153 including 114 forest biome species (63%) (Ntiamoa-Baidu et al. 2001).

Mist-netting at Boi-Tano was carried out in mature secondary forest from 30 October – 1 November within compartment 63, part of the Globally Significant Biodiversity Area (GSBA). Birds trapped at this site included *Bleda eximia*, *Illadopsis rufescens* and other species such as *Alcedo leucogaster*, *Indicator maculates*, *Smithornis rufolateralis* and *Criniger barbatus* that are typical of mature forest. Nineteen species were trapped at a rate of 2.5 individuals per 100 meter net hours (see Appendix 9).

In addition, a number of species were observed at Boi-Tano that are rare and poorly known in either Ghana or Upper Guinea. These include *Spizaetus africanus*, *Accipiter melanoleucus*, *Chrysococcyx flavigularis*, *Otus icterorhynchus*, *Rhaphidura sabini*, *Megabyas flammulatus*, *Pholidornis rushiae*, *Cinnyris johannae* and *C. batesi*. Breeding activity of only two species was observed (see Appendix 8).

## Krokosua Hills

We recorded 138 species over seven days in this forest (see Appendix 8) amongst which were four species of global conservation concern: *Bycanistes cylindricus*, *Bleda eximia*, *Illadopsis rufescens* and *Lamprotornis cupreocauda* (BirdLife International 2000; Table 6.1). These four species plus *Apalis sharpii* are all restricted-range species endemic to the Upper Guinea forests. We observed 97 Guinea-Congo forest biome species, 54% of the 180 forest biome species known from Ghana (Ntiamoa-Baidu et al. 2001). This is a large component of the forest-restricted species in the country. The total

**Table 6.2.** Number of Guinea-Congo Forests biome species recorded from protected areas in Ghana (Ntiamoah-Baidu et al. 2001, this study). Sites surveyed in this study in bold.

| Rank | Site | No. of forest biome species |
|------|------|-----------------------------|
| 1 | Kakum NP - Assin Atandaso RR | 137 |
| 2 | **Boi-Tano FR** | 114 |
| 3 | Bia NP and RR | 114 |
| 4 | **Draw River FR** | 111 |
| 5 | Ankasa RR - Nini Suhien NP | 106 |
| 6 | **Krokosua FR** | 102 |
| 7 | Tano-Ehuro FR | 95 |
| 8 | Boin River FR | 88 |
| 9 | Yoyo FR | 79 |
| 10 | Mamiri FR | 78 |

**Table 6.1.** Species of global conservation concern recorded at each site. F-frequent, U-uncommon, R-rare, x-previously recorded. See Appendix 8 for further explanation of status codes.

| | Species | Sites | | | Threat Status |
|---|---------|-------|---|---|---------------|
| | | **Draw River** | **Boi-Tano** | **Krokosua** | |
| 1 | *Tigriornis leucolophus* | | x | | DD |
| 2 | *Agelastes meleagrides* | | x | | VU |
| 3 | *Bycanistes cylindricus* | x | | R | NT |
| 4 | *Ceratogymna elata* | x | | | NT |
| 5 | *Bleda eximia* | U | U | R | VU |
| 6 | *Criniger olivaceus* | R | x | | VU |
| 7 | *Illadopsis rufescens* | x | U | R | NT |
| 8 | *Lamprotornis cupreocauda* | F | x | U | NT |
| | Total species | 6 | 6 | 4 | |

Threat status (BirdLife International 2000):

VU = Vulnerable: species facing a high risk of extinction in the medium-term future.

DD = Data Deficient: species for which there is inadequate information to make an assessment of its risk of extinction.

NT = Near Threatened: species coming very close to qualifying as Vulnerable.

number of bird species now known from Krokosua is 146 including 102 forest biome species (57%) (A. Asamoah pers. obs.).

Mist-netting at Krokosua was carried out at two sites, the first from 5-6 November and the second from 8-9 November. Forest condition was generally non-uniform as several areas had been degraded by logging, fires and farming. Mist-netting was carried out in patches where the forest quality was comparatively good. The assemblage of birds trapped was dominated by species associated with secondary forest growth but also included species that are typical of mature forest, including *Bleda syndactyla* and *Phyllastrephus albigularis*, the latter a species of semi-deciduous forest. Fourteen species were trapped at a rate of 1.4 per 100 meter net hours (see Appendix 9).

In addition, a number of species were observed at Krokosua that are rare and poorly known in either Ghana or Upper Guinea. These include *Spizaetus africanus, Otus icterorhynchus, Bubo poensis, Bubo leucostictus, Glaucidium tephronotum, Rhaphidura sabini, Telecanthura melanopygia, Neafrapus cassini, Sylvietta denti, Megabyas flammulatus, Poeoptera lugubris* and *Ploceus preussi*. Breeding activity of five species was observed (see Appendix 8).

### Notes on specific species
See Table 6.1 for explanation of threat status. Status in West Africa from Borrow and Demey (2001). Recent status in Ghana and IBA details are from Ntiamoa-Baidu (2001). Reference to Grimes (1987) is made where there is little recent data on an individual species.

### Species of conservation concern
*Bycanistes cylindricus* Brown-cheeked Hornbill (NT). One observed in Krokosua Hills by T. Larsen (pers. comm.) along a tributary of the Bia River close to the road entering the forest. This was the only large hornbill observed during our survey of the three reserves. An uncommon to locally fairly common forest resident and Upper Guinea endemic. Less common than in the past in Ghana (Grimes 1987).

*Bleda eximia* Green-tailed Bristlebill (VU). One individual was trapped at both Draw River and Boi-Tano. Additionally this species was seen and heard at a different location in Draw River and also heard at both Boi-Tano and Krokosua Hills at different locations within 3 km of the camps. This species was not previously known from Boi-Tano or Krokosua Hills. Rare forest resident and Upper Guinea endemic.

*Criniger olivaceus* Yellow-bearded Greenbul (VU). Two individuals were tape-recorded and subsequently seen 3 km north-west of the camp in Draw River. This species was previously known from this site. Generally rare forest resident and Upper Guinea endemic.

*Illadopsis rufescens* Rufous-winged Illadopsis (NT). One individual was trapped in Boi-Tano and one was also heard singing close by on a subsequent day. Individuals were heard singing at two different locations on one day in Krokosua Hills. Uncommon to rare forest resident and Upper Guinea endemic.

*Lamprotornis cupreocauda* Copper-tailed Glossy Starling (NT). Seen on four days in Draw River including one breeding pair. Seen on two days in Krokosua Hills close to the plantations of Mmem village. Fairly common to locally common forest resident and Upper Guinea endemic.

### Other rare or poorly known species
*Dryotriorchis spectabilis* Congo Serpent Eagle. Seen at two different locations on separate days in Draw River. Not previously known from any IBA in Ghana. Very few recent records (Grimes 1987). Scarce to locally fairly common in forest.

*Accipiter erythropus* Red-thighed Sparrowhawk. One seen just outside the Draw River boundary. There are relatively few recent records of this uncommon to scarce forest resident (Grimes 1987, Ntiamoa-Baidu et al. 2001).

*A. melanoleucus* Black Sparrowhawk. One sighting in both Draw River and Boi-Tano. Previously known from Draw River. Very few recent records from Ghana where it may be overlooked (Grimes 1987). A fairly common to rare species in West Africa.

*Spizaetus africanus* Cassin's Hawk Eagle. Seen twice in Boi-Tano (J. McCullough pers. comm.) and once in Krokosua Hills. A scarce to locally fairly common forest resident.

*Chrysococcyx flavigularis* Yellow-throated Cuckoo. Seen once in Boi-Tano. Known from few sites in Ghana although previously recorded from Boi-Tano. Rare in Upper Guinea.

*Otus icterorhynchus* Sandy Scops Owl. Recorded calling at all sites. In Ghana, recently recorded only from Kakum National Park (NP). No other records since the 1950s (Grimes 1987). Uncommon to rare forest resident.

*Bubo poensis* Fraser's Eagle Owl. Several individuals at different locations were recorded calling at Draw River and Krokosua Hills. Recent records are only from Kakum NP in Ghana although Grimes (1987) thought it uncommon in Ghana. An uncommon to fairly common forest resident.

*Bubo leucostictus* Akun Eagle Owl. One individual was recorded calling at Krokosua Hills on two nights on the plantation-forest boundary at Mmem village. Previously known only from Kakum NP in Ghana and from old skins (Grimes 1987, Ntiamoa-Baidu et al. 2001). A rare to locally fairly common forest resident.

*Glaucidium tephronotum* Red-chested Owlet. A maximum of three individuals were recorded calling at Krokosua Hills on two nights. Known previously from four IBAs and from old skins in Ghana (Grimes 1987). A rare to uncommon forest resident.

*Rhaphidura sabini* Sabine's Spinetail. This species was fairly common in Draw River and was seen once in Boi-Tano and twice in Krokosua Hills. Known from only five IBAs although Grimes (1987) describes it as 'not uncommon'. A locally common to uncommon forest resident.

*Telacanthura melanopygia* Black Spinetail. Two individuals seen close to the forest edge in Krokosua Hills. Recent records are only from Kakum NP in Ghana although Grimes (1987) thought it not uncommon. A rare to locally uncommon forest resident.

*Neafrapus cassini* Cassin's Spinetail. Two individuals seen in Draw River and one in Krokosua Hills. Some scattered records throughout the forest zone in Ghana (Grimes 1987, Ntiamoa-Baidu et al. 2001). Locally fairly common in forest.

*Prodotiscus insignis* Cassin's Honeybird. Seen once in Draw River. Two possible sightings in Krokosua Hills. A rare species in Ghana although recently found more frequently (Grimes 1987, Ntiamoa-Baidu et al. 2001). Scarce to uncommon resident in West Africa.

*Sylvietta denti* Lemon-bellied Crombec. Recorded on three days with a maximum of five individuals in Krokosua Hills. Known from only two other sites in Ghana. Rare in Ghana (Grimes 1987) but fairly common elsewhere in West African forests.

*Megabyas flammulatus* Shrike-flycatcher. A pair recorded at Boi-Tano and a female at Krokosua Hills. Recently known only from Kakum NP with only some older records from elsewhere in Ghana (Grimes 1987, Ntiamoa-Baidu et al. 2001). An uncommon forest resident in West Africa.

*Cinnyris johannae* Johanna's Sunbird. One observation in Draw River and two in Boi-Tano. Recorded recently from Ankasa, Nini Suhien and Boi-Tano but overall rare in Ghana (Grimes 1987, Ntiamoa-Baidu et al. 2001). A rare to locally fairly common forest resident in West Africa.

*Cinnyris batesi* Bates's Sunbird. One group seen in each of Draw River and Boi-Tano. Previously unrecorded from any IBA in Ghana but possibly overlooked (Grimes 1987). Rare to locally common in West African forests.

*Poeoptera lugubris* Narrow-tailed Starling. A group of six and a lone individual observed in Krokosua Hills. Recently known only from Kakum NP in Ghana. Fairly common to scarce and rather local in West African forests.

*Ploceus preussi* Preuss's Golden-backed Weaver. Two seen close to the forest boundary in Krokosua Hills. Scarce to uncommon in West African forests.

*Parmoptila rubrifrons* Red-fronted Antpecker. A pair trapped over a river in Draw River. A male seen in a mixed-species flock close to this location. Rare to locally fairly common in West Africa.

## DISCUSSION

The total number of 170 species recorded across all sites is relatively high in view of the short study period and compares well with the 179 species recorded in Haute Dodo and Cavally Forest reserves, Côte d'Ivoire over a similar period (Demey and Rainey 2005). These three sites are now all in the top six forests in Ghana in terms of the number of Guinea-Congo Forests biome species recorded from each site (see Table 6.2). This gives an indication of the high quality of the forests.

### Draw River
We recorded three species of global conservation concern (category A1) (see Table 6.1) at this site and four restricted-range species (category A2) (see Fishpool and Evans 2001 for category definitions). Of the 146 species now known from Draw River, 111 (62% of the Ghana total) are Guinea-Congo Forests biome species (category A3). In total, six species of conservation concern are now known from Draw River and together our observations have increased the importance of this site as an IBA (Fishpool and Evans 2001). Remarkably, it can be seen in Table 6.2 that more forest biome species have been recorded in Draw River than in the adjacent combined protected areas of Ankasa Resource Reserve (RR) and Nini Suhien NP. This is despite Ankasa and Nini Suhien being larger, 'well-studied' and supporting the largest number of threatened bird species of any forest site in Ghana (Ntiamoa-Baidu et al. 2001). Given this, consideration should be given to upgrading the protection status of Draw River (see recommendations below). Further surveys in different habitats (e.g. wetlands and waterways) may find many more species at this site.

### Boi-Tano
We recorded two species of global conservation concern at this site (category A1) (see Table 6.1) and three restricted-range species (category A2). Of the 153 species now known from Boi-Tano, 114 (63% of the Ghana total) are Guinea-Congo Forests biome species (category A3). In total, six species of conservation concern are now known from Boi-Tano and together these have increased the importance of this site as an IBA (Fishpool and Evans 2001). In Ghana, this site now ranks second only to Kakum NP in terms of the number of forest biome species recorded (Table 6.2). Previous surveys have found the Vulnerable *Agelastes meleagrides* in the reserve (Dutson and Branscombe 1990). As this large

species is one of the most threatened in Ghana, as this forest has so many forest biome species and as only relatively small sections of Boi-Tano and the adjacent Tano Nimiri have fully protected habitat, consideration should be given to upgrading the protection status of Boi-Tano. Higher status could provide better protection against hunting. Development of a corridor between this site and Nini Suhien NP to the south could benefit many species by increasing the total area of good quality habitat available to them.

## Krokosua Hills

We recorded four species of global conservation concern (category A1) at this site and five restricted-range species (category A2) (see Fishpool and Evans 2001 for category definitions). Of the 146 species recorded from Krokosua Hills, 102 (57% of the Ghana total) were Guinea-Congo Forests biome species (category A3). Together these meet three of the criteria for selection of the site as an IBA (Fishpool and Evans 2001). This site therefore is of national and international importance for the conservation of bird biodiversity. Recommendations relating to this site are given below. We did not find some species typical of semi-deciduous forest and hills such as *Pyrrhurus scandens*, *Phyllanthus atripennis* and *Dicrurus ludwigii* that you might expect in such habitat. It is possible that further surveys may locate them but it is also possible that the altitudinal range in Krokosua Hills is not great enough for some of these species.

*Bleda eximia* was recorded at each site and this is a good indicator of forest quality. Although few other species of global conservation concern were recorded many of these are only found at low density and may be recorded during future surveys. Indeed, a number of other such species have already been recorded from Draw River and Boi-Tano (Ntiamoa-Baidu et al. 2001). Relatively few forest edge or farmbush species were recorded at each site and this is a further indicator of the intact status of these forests.

Large species such as guineafowl, large birds of prey, *Corythaeola cristata* and large hornbills were rarely recorded during these surveys. Our experience of guineafowl at other sites is that the feathers of Crested Guineafowl *Guttera pucherani* are commonly found even if the birds are not often seen. Conversations with hunters in Krokosua Hills indicated that they did not know *Agelastes meleagrides* and had not seen *Guttera pucherani* for four years or *Corythaeola cristata* for two years. HJR's experience of hornbills in Côte d'Ivoire and their movements between forests is that they are most common in southern evergreen forests in the second half of the year and more common in semi-deciduous forests in the first half of the year. As Draw River (wet evergreen forest), Boi-Tano (moist evergreen) and Krokosua Hills (semi-deciduous) cover all these forest types it is remarkable that we did not encounter them in significant numbers in at least one of the sites. Hornbills and *Corythaeola cristata* are loud, vocal birds and would normally be recorded during a survey of this type by their calls. It is possible that the populations of these and other large species have been reduced by hunt-

ing. Holbech (1996) provides evidence that large hornbills are being hunted more frequently in Ghana in recent years than previously and that this may be related to the decline in abundance of other large species. He also shows that they are hunted more often in smaller forests with higher levels of logging taking place. Thus fragmentation and ease of access may also have increased hunting rates of these species (Holbech 2005). The high density in Draw River and Boi-Tano of the relatively small *Bycanistes fistulator* may be related to reduced competition from larger hornbills. Given the importance of some of these species in fruit dispersal and therefore forest tree regeneration (Whitney et al. 1998), surveys for presence and abundance of these large species would be of considerable value. Sites from which they are known to have been abundant (e.g. Ankasa RR and Bia NP and RR) should be priorities for such surveys.

## CONSERVATION RECOMMENDATIONS

Considering the high conservation value of all three forests, we recommend the following:

1.  Draw River forest reserve should have its status raised to receive the highest level of protection as found in Ankasa RR and Nini Suhien NP.

2.  Hunting at all sites, and, in particular of threatened species, should be monitored and regulations enforced.

3.  Boi-Tano lies only a short distance from the northern edge of Nini Suhien NP. Consideration should be given to linking the two protected areas with a corridor of well-protected forest along natural corridors such as rivers. This should be wide enough to allow movements of a range of species. This could be important for a number of bird species that make long distance movements such as hornbills and parrots as well as for large mammals such as elephants. It would also decrease the vulnerability of other species to stochastic extinction events and facilitate their dispersal.

4.  Boi-Tano should receive the highest protection status, particularly if linked to Nini Suhien NP, as it is one of the few known sites for a number of threatened species including *Agelastes meleagrides*.

5.  Krokosua Hills should receive higher protected area status in view of the quality of the birds we found there in a relatively short time.

6.  The boundaries of the farmland of villages within the forest reserves should be as clearly demarcated as they have been for the external boundaries. These boundaries within the forests should be enforced.

7. Large bird species should be the focus of further surveys in areas where they were previously known to be relatively abundant. Similarly, the three forest reserves that we visited should be surveyed during the first half of the year to assess hornbill abundance. This would provide an indication of the status of these species in Ghana.

8. The GSBAs in adjoining sites (e.g. Boi-Tano and Tano Nimiri) should be linked by sufficiently wide corridors of good quality forest to enable dispersal of wildlife. Likewise, timber extraction at each site should be carried out at lower intensity to conserve forest quality.

9. Programs should be put in place to support local people to develop alternative sources of protein. Local communities should also be educated not to hunt threatened species and encouraged not to use snares that are indiscriminate and often forgotten.

## REFERENCES

Allport, G.A., M. Ausden, P.V. Hayman, P. Robertson and P. Wood. 1989. The conservation of the birds of Gola Forest, Sierra Leone. ICBP Study Report No. 38. International Council for Bird Preservation. Cambridge, UK.

Bakarr, M., B. Bailey, D. Byler, R. Ham, S. Olivieri and M. Omland. 2001. From the Forest to the Sea: Biodiversity Connections from Guinea to Togo. Map: Biodiversity Priorities from Guinea to Togo. Conservation International. Washington, DC.

BirdLife International. 2000. Threatened Birds of the World. Lynx Edicions and BirdLife International. Barcelona, Spain and Cambridge, UK.

Borrow, N. and R. Demey. 2001. Birds of Western Africa. Christopher Helm. London.

Demey, R. and Rainey, H.J. 2005. A rapid survey of the birds of Haute Dodo and Cavally Forest reserves. *In*: Lauginie, F., G. Rondeau and L.E. Alonso (eds.). A biological assessment of two classified forests in Southwestern Côte d'Ivoire. 34. Conservation International. Washington, DC, RAP Bulletin of Biological Assessment.

Demey, R. and Rainey, H.J. 2004. A preliminary survey of the birds of the Forêt Classée du Pic de Fon. *In*: McCullough, J. (ed.). A biological assessment of the terrestrial ecosystems of the Forêt Classée du Pic de Fon, Simandou Range, Guinea. RAP Bulletin of Biological Assessment 35. Conservation International. Washington, DC. Pp. 63-68.

Dutson, G. and J. Branscombe. 1990. Rainforest birds in southwest Ghana. Results of the ornithological work of the Cambridge-Ghana Rainforest Project 1988 and the Ghana Rainforest Expedition '89. ICBP Study Report 46. International Council for Bird Preservation. Cambridge, UK.

Fishpool, L.D.C. and M.I. Evans (eds.). 2001. Important Bird Areas in Africa and Associated Islands: Priority sites for conservation. Pisces Publications and BirdLife International, Newbury and Cambridge, UK.

Gartshore, M.E., P.D. Taylor and I.S. Francis. 1995. Forest Birds in Côte d'Ivoire. A survey of Taï National Park and other forests and forestry plantations, 1989–1991. Study Report No. 58. BirdLife International. Cambridge, UK.

Grimes, L.G. 1987. The birds of Ghana. BOU Checklist No. 9. British Ornithologists' Union. London.

Hawthorne, W.D. and M. Abu-Juam. 1995. Forest Protection in Ghana. IUCN/ODA/Forest Department. Gland, Switzerland, Cambridge, UK and Kumasi, Ghana.

Holbech, L.H. 2005. The implications of selective logging and forest fragmentation for the conservation of avian diversity in evergreen forests of south-west Ghana. Bird Conservation International, 15: 27-52.

Holbech, L.H. 1996. Faunistic diversity and game production contra human activities in the Ghana high forest zone, with reference to the Western Region. Unpublished Ph.D. thesis. Copenhagen, Denmark. University of Copenhagen.

ICBP. 1992. Putting biodiversity on the map: priority areas for global conservation. International Council for Bird Preservation. Cambridge, UK.

Ntiamoa-Baidu, Y., E.H. Owusu, D.T. Daramani and A.A. Nuoh. 2001. Ghana. *In*: Fishpool, L.D.C. and M.I. Evans (eds.). Important Bird Areas in Africa and Associated Islands: Priority sites for conservation. Newbury and Cambridge, UK: Pisces Publications and BirdLife International. Pp. 367-389.

Whitney, K.D., M.K. Fogiel, A.M. Lamperti, K.M. Holbrook, D.J. Stauffer, B.D. Hardesty, V.T. Parker and T.B. Smith. 1998. Seed dispersal by *Ceratogymna* hornbills in the Dja Reserve, Cameroon. Journal of Tropical Ecology. 14: 351-371.

# Chapter 7

## Rapid Assessment of Small Mammals at Draw River, Boi-Tano, and Krokosua Hills

*Jan Decher, James Oppong and Jakob Fahr*

## SUMMARY

This RAP provided a unique opportunity to increase our knowledge of the small mammal diversity of southwestern Ghana. 105 terrestrial small mammal captures, composed of six species of shrews and ten species of rodents, and 82 bat captures composed of 15 species were made. The shrew *Crocidura obscurior* is the first record for Ghana. *Crocidura buettikoferi* and *C. foxi* are first records for southwestern Ghana. At Krokosua Hills we documented the rare microbat *Scotophilus nucella* (IUCN Red List: Vulnerable), which was described in 1984 and hitherto known from only ten specimens from Ghana, Côte d'Ivoire and Uganda. Overall the small mammal species composition clearly reflects a forest fauna. Especially among the bats sampled, not a single savannah species was present despite partially degraded forest conditions.

## INTRODUCTION

During the West Africa Conservation Priority Setting Workshop in 1999 (Bakarr et al. 2001), the region of eastern Côte d'Ivoire and western Ghana, including all three forest reserves sampled during this RAP, was designated a high priority area for mammal protection and was given an "exceptionally high" priority in the final ranking. On the Ghanaian side of this region most of the original vegetation cover is only preserved in a patchwork of forest reserves and two national parks (Ankasa Resource Reserve with Nini Suhien National Park and Bia National Park). Only a few small mammal studies have been conducted in this area and, with one exception (Holbech 1999), most studies are more than 25 years old (Cole 1975, Jeffrey 1973, 1975, 1977). Reliable surveys enhancing our knowledge of the current small mammal distributions and their conservation status throughout Ghana are urgently needed. Small mammal diversity of a given area cannot be exhaustively assessed within the short period of 3-5 days of sampling per site typical for a Rapid Assessment Program (RAP) survey. Small mammal trapping results from these RAPs can nevertheless indicate how much of a forest community of small mammals still exists, or to what extent savanna and farmbush species have already invaded isolated forests or forest edges (Decher and Abedi-Lartey 2002, Decher 2004, Decher et al. 2005) and they may on occasion yield surprising new geographic records (Fahr and Ebigbo 2003).

The most complete review of Ghanaian mammal distributions to date is that by Grubb et al. (1999). For the southwestern part of Ghana, originally covered with rainforest, these authors provide a conservative estimate of 8 shrew, 60 bat, and 29 rodent species, including several savannah species that have recently invaded this nowadays highly fragmented and disturbed forest region.

## MATERIALS AND METHODS

Small mammal survey techniques followed those described by Voss and Emmons (1996) and Martin et al. (2001) and comply with recommended guidelines and standard methods for mammal field work (Animal Care and Use Committee 1998, Wilson et al. 1996).

At all three sites in the Western Region of Ghana, we (JD and JO) installed two or three trap lines, with 40 traps each, consisting of standard Sherman live traps except at the first site (Draw River), where we also employed a few Victor rat and Museum Special snap traps. In addition, at each site three pitfall lines with driftfence and at least six pitfall buckets each were employed and shared with the herpetology team. Sherman traps and snap traps were baited with a mixture of fresh palmnut shavings and rolled oats. At Draw River, five Tomahawk traps were loaned to local hunters to obtain larger species of small mammals. Trap and pitfall lines were checked every morning and captured animals were measured, identified, and released at the capture site, or kept as voucher specimens for further identification. Field identifications were based mainly on Hutterer and Happold (1983) and Meester and Setzer (1971). Voucher specimens were deposited at the Museum Alexander Koenig, Bonn, Germany (shrews and bats) and the United States National Museum, Washington, DC (rodents; see Appendix 10). For each captured animal some microhabitat data including percentage of canopy cover, distance to nearest tree and nearest fallen log, trap height, slope, and percentage of five groundcover types were recorded. Bats were sampled with 2-8 net units (1 unit = one 6 x 2.5 m net) for each site and night, placed in locations considered as suitable flyways or over water. Nets were usually open between 18:00 h and 21:30 h. Net locations were usually in walking distance (500-1000 m) from the camp sites with one location at Draw River being about 5 km distant.

Species accumulation curves were generated using the program EstimateS, Version 7, with 100 randomization runs (Colwell 2004). The IUCN Red List status is based on the recent update available at www.redlist.org, following the Global Mammal Assessment (GMA) of African small mammals in January 2004 (IUCN 2004).

Coordinates for the camp sites at each locality, obtained with a Garmin GPS 12 receiving unit, are as follows:
Site 1 (Draw River):     5°12'N  2°24'W
Site 2 (Boi-Tano):       5°32'N  2°37'W
Site 3 (Krokosua Hills): 6°37'N  2°51'W

## RESULTS AND DISCUSSION

Table 7.1 shows the results of terrestrial small mammal trapping in all three sampled reserves. Trap effort encompassed a total of 1210 trapnights (= number of traps + buckets x number of nights trapped). 105 individual captures were made, representing six species of shrews and ten species of rodents, the latter including two Giant Rats (*Cricetomys emini*) captured by locals, and one individual of Edward's Swamp Rat (*Malacomys edwardsi*) found dead in a snare. Daily trapping success ranged from 3.1% to 20.7% for single sites, with an overall trapping success of 8.4%.

### Shrews

Draw River. The shrew *Crocidura buettikoferi* was captured only at this site, as was *Crocidura foxi*. The capture of *Crocidura buettikoferi* is the second or third record of this species from Ghana. This species was previously captured by one of us (JD) at Adumanya Sacred Grove on the Accra Plains (Decher et al. 1997) and Grubb et al. (1999) mention two additional but unverified British Museum specimens from Wenchi in the Brong Ahafo Region. *Crocidura foxi* was not mentioned by Grubb et al. (1999) for Ghana, but two specimens have since been caught in a forest remnant at Apesokubi in the Volta Region (Decher and Abedi-Lartey 2002). This shrew was described as inhabiting "derived and Guinea savannah" in Nigeria (Hutterer and Happold 1983) but it appears to have a much broader habitat preference.

Boi Tano. Two specimens of the climbing mouse-tailed shrew, *Crocidura muricauda*, a forest species, were captured at this site. The species was listed as "expected" for adjacent Ankasa Resource Reserve and Nini Suhien National Park by Holbech (1999) under the name *C. dolichura*. According to Hutterer (1993), *C. muricauda* can be considered a species distinct from *C. dolichura* and is an Upper Guinea endemic, first reported from Mount Coffee, Liberia (Miller 1900). Grubb et al. (1999) map two localities in southwestern Ghana (near Krokosua Hills) but give no reference. This is a typical forest shrew and was shown to be the second most common shrew (after *C. obscurior*) in Taï National Park, Côte d'Ivoire, using intensive pitfall trapping (Barriere et al. in litt.).

Krokosua Hills. Only at Krokosua Hills did we capture one specimen of the large *Crocidura olivieri*, which is a common shrew throughout Ghana (Grubb et al. 1999). One shrew assigned to the *C. crossei* / *C. jouvenetae*-species complex was caught at each Boi-Tano and Krokosua Hills. The small rain forest shrew *Crocidura obscurior* was not mentioned by Grubb et al. (1999) for Ghana and our records from all three sites appear to be the first for Ghana (R. Hutterer in litt.).

### Rodents

Draw River. Seven species of rodents were verified for Draw River out of 27 captures in 386 trapnights. Daily trap success for all terrestrial mammals ranged from 7.1% to 9.4% (mean 8.0%). Additionally, one Giant Rat (*Cricetomys emini*) was captured in farmbush adjacent to the forest in a Tomahawk trap loaned to local hunters. The most common species caught was *Hylomyscus alleni*, outnumbering all other species added together and frequently caught in traps set above ground on branches or stilt roots. One specimen each of the eastern Upper Guinea endemic *Malacomys cansdalei* (Cansdale's Swamp Rat) and the Upper Guinea endemic *Dephomys defua* (Defua Rat) were captured at this site. *Dephomys defua* is known from several localities of the Ghanaian forest zone but is uncommonly collected (Grubb et al. 1999).

**Table 7.1:** Non-volant small mammal captures at Draw River, Boi-Tano and Krokosua Hills forest reserves, Ghana, 24 Oct.- 9 Nov. 2003

| Localities: | Draw River (DR) | | | | | Boi-Tano (BT) | | | | | Krokosua Hills (KH) | | | | | | Grand Total |
|---|---|---|---|---|---|---|---|---|---|---|---|---|---|---|---|---|---|
| Date: | 24 Oct | 25 Oct | 26 Oct | 27 Oct | Total DR | 29 Oct | 30 Oct | 31 Oct | 1 Nov | Total BT | 5 Nov | 6 Nov | 7 Nov | 8 Nov | 9 Nov | Total KH | Grand Total |
| **Order/Species** | | | | | | | | | | | | | | | | | |
| **INSECTIVORA** | | | | | | | | | | | | | | | | | |
| *Crocidura buettikoferi* | | | 1 | | 1 | | | | | | | | | | | | 1 |
| *Crocidura crosseil jouventae* | | | | | | | | 1 | | 1 | | | | | | 1 | 2 |
| *Crocidura foxi* | | 1 | | 2 | 3 | | | | | | | | | | | | 3 |
| *Crocidura muricauda* | | | | | | | 1 | | 1 | 2 | | | | | | | 2 |
| *Crocidura obscurior* | 1 | | | | 1 | | | 1 | 1 | 2 | | 1 | 1 | 1 | | 4 | 7 |
| *Crocidura olivieri* | | | | | | | | | | | | | | | 1 | 1 | 1 |
| **RODENTIA** | | | | | | | | | | | | | | | | | |
| *Cricetomys emini* | | (1) | | | (1) | (1) | | | | (1) | | | | | | | (2)* |
| *Dephomys defua* | | | 1 | | 1 | | | | 1 | 1 | | | | | | | 2 |
| *Grammomys rutilans* | | | | | | | | | 1 | 1 | | | | | | | 1 |
| *Graphiurus nagtglasii* | | | | | | | | | | | | | 1 | | | 1 | 1 |
| *Hybomys trivirgatus* | | | 1 | | 1 | | 1 | 1 | | 2 | | | | | | | 3 |
| *Hylomyscus alleni* | | 4 | 6 | 8 | 18 | | 1 | 4 | 4 | 9 | 1 | 3 | 6 | 11 | 5 | 26 | 53 |
| *Lophuromys sikapusi* | | 1 | | 2 | 3 | | | | 1 | 1 | | | | | | | 4 |
| *Malacomys cansdalei* | | | 1 | | 1 | | | | | | | | | | | | 1 |
| *Malacomys edwardsi* | | | | | | (1) | | | | (1) | | 1 | 1 | 1 | | 3 | 3(4)* |
| *Praomys tullbergi* | | 1 | | 1 | 2 | | | 2 | | 2 | 2 | 1 | 3 | 3 | 5 | 14 | 18 |
| Total No. of Captures: | 1 | 7 | 10 | 13 | 31 | n/a | 3 | 8 | 10 | 21 | 3 | 6 | 13 | 16 | 12 | 50 | 102(105)* |
| No. of Species: | 1 | 5 | 5 | 4 | 10 | 2 | 3 | 4 | 7 | 11 | 2 | 4 | 6 | 4 | 4 | 7 | 16 |
| Triggered Traps: | 0 | 80 | 120 | 120 | - | n/a | 80 | 80 | 80 | - | 80 | 80 | 120 | 120 | 40 | - | - |
| Pitfalls: | 12 | 18 | 18 | 18 | - | n/a | 18 | 18 | 18 | - | 18 | 18 | 18 | 18 | 18 | - | - |
| Total No. of Traps/ trapnights: | 12 | 98 | 138 | 138 | 386 | n/a | 98 | 98 | 98 | 294 | 98 | 98 | 138 | 138 | 58 | 530 | 1210 |
| Trap Success (%): | 8.3 | 7.1 | 7.2 | 9.4 | 8.0 | n/a | 3.1 | 8.2 | 10.2 | 7.1 | 3.1 | 6.1 | 9.4 | 11.6 | 20.7 | 9.4 | 8.4 |

* = Numbers in parentheses indicate specimens that were obtained by methods other than trap lines (not included in calculation of species accumulation curve; Fig. 3a).

A biological assessment of the terrestrial ecosystems of the Draw River, Boi-Tano,
Tano Nimiri and Krokosua Hills forest reserves, southwestern Ghana

59

Boi-Tano. With 16 captures and eight species of rodents in 294 trapnights, the species distribution pattern in general was quite similar to Draw River. Daily trap success for all terrestrial mammals ranged from 3.1% to 10.2% (mean 7.1%). *Hylomyscus alleni* was again the most dominant rodent. Another specimen of *Dephomys defua* was collected here, as well as the only specimen of the Shining Thicket Rat (*Grammomys rutilans*), which was caught in a trap set on vines at 1.8 m above ground, confirming this species' arboreal habits (Jeffrey 1973, 1977). Another specimen of the Giant Rat (*Cricetomys emini*) was received from local hunters at this location captured using local methods and one individual of Edward's Swamp Rat (*Malacomys edwardsi*) was found dead in a snare.

Krokosua Hills. At Krokosua Hills forest reserve individuals from only four species of rodents made up 44 captures, even though total trapping effort was greater (530 trapnights). Mean daily trap success for all terrestrial mammals was 9.4% and ranged from 3.1% to 20.7%, the latter on the last morning in a trapline set along a narrow forest trail. *Hylomyscus alleni* was again the most frequently captured species (26 individuals) with *Praomys tullbergi* ranking second (14 individuals). One specimen of the dormouse *Graphiurus nagtglasii* was captured at the entrance of a hole at the bottom of a large buttress tree. The use of hollow trees and this rodent's frequent association with roosting *Hipposideros cyclops* (a bat species that was captured in a nearby net) has been previously mentioned in the literature (Schlitter et al. 1985). Other *G. nagtglasii* were observed repeatedly at night along the access road to the village of Mmem foraging in a

flowering *Solanum erianthum* bush. We follow Grubb and Ansell (1996) in using the name *G. nagtglasii* for *G. hueti*.

If we compare our rodent results at Krokosua Hills to a much more extensive study conducted in the same area in 1970 (Jeffrey 1977; see Table 7.2), our results for primary forest are qualitatively relatively similar to the earlier findings, but missing both *Hybomys trivirgatus* and *Lophuromys sikapusi*, common in the earlier study. Quantitatively, our study yielded rather different proportions than the earlier work, for example a much higher abundance of *Hylomyscus alleni* (Table 7.2). The absence of *H. trivirgatus* and *L. sikapusi* might be explained by our short sampling period. However, both species (and *Dephomys defua*) were recorded during our much shorter sampling periods at Draw River and Boi-Tano. *H. trivirgatus* seems to be a more obligate forest species than *L. sikapusi* as shown by data from Jeffrey (1975, 1977; see Table 7.2), who caught *L. sikapusi* in the Krokosua Hills area abundantly in natural clearings and in all man-made habitat with grass cover as the second most common species after *Praomys tullbergi*. Other studies also characterize *L. sikapusi* as being abundant in more open grassy areas (Dieterlen 1976) and even in 4-5 year old palm plantations (Bellier 1965). However, the present survey yielded specimens of *L. sikapusi* and *H. trivirgatus* only from the "wet evergreen" (Draw River and Boi-Tano) forest type, using the designation by Hall and Swaine (1981; see also map "Priority areas for plants" in Bakarr et al. 2001:32). Similar to our present findings, other recent surveys confirmed the presence of both *H. trivirgatus* and *L. sikapusi* in the "wet evergreen" Haute Dodo and Cavally forest reserves of Côte d'Ivoire (Decher et al. 2005), and their absence or scarcity

**Table 7.2:** RAP survey rodent results at Krokosua Hills compared to 1970 results from Jeffrey (1977)

| | 2003 study (530 trapnights) | 1970 study (27 469 trapnights) | |
|---|---|---|---|
| | Primary Forest * | Primary Forest (Krokosua & Bia Tribut. North F.R.) | Natural Clearings at Krokosua |
| *Mus musculoides* | 0 | 0 | 2 (1.8 %) |
| *Lophuromys sikapusi* | 0 | 9 (4.6 %) | 72 (65.5 %) |
| *Malacomys edwardsi* | 4 (8.9 %) | 28 (14.4 %) | 0 |
| *Lemniscomys striatus* | 0 | 0 | 1 (0.9 %) |
| *Grammomys rutilans* (= *Thamnomys rutilans* in Jeffrey 1977) | 0 | 0 | 1 (0.9 %) |
| *Hybomys trivirgatus* | 0 | 39 (20.0 %) | 4 (3.6 %) |
| *Praomys tullbergi* | 14 (31.1 %) | 111 (56.9 %) | 24 (21.8 %) |
| *Dephomys defua* | 0 | 0 | 4 (3.6 %) |
| *Hylomyscus alleni* | 26 (57.8 %) | 7 (3.6 %) | 0 |
| *Graphiurus lorraineus* (=*G. murinus* in Jeffrey 1977) | 0 | 0 | 2 (1.8 %) |
| *Graphiurus nagtglasii* (= *G. hueti* in Jeffrey 1977) | 1 (2.2 %) | 1 (0.5 %) | 0 |

* last selective logging in 1991 according to Hawthorne and Abu-Juam (1995).

in the "dry semideciduous" forests of the Volta Region of Ghana (Decher and Abedi Lartey 2002), and on the Pic de Fon Mountain in Guinea (Decher 2004). Perhaps these species are now less common at Krokosua Hills due to the unprecedented forest losses and abundant canopy disturbances in comparison to overall conditions thirty years ago (Jeffrey 1975, 1977), possibly resulting in increased aridity of the region and thus limiting food sources for these species which are known to be predominantly insectivorous (Cole 1975).

The high abundance of *Hylomyscus alleni* at all sites during this study may be a result of our effort to set more traps above ground level on branches and vines, where this arboreal species is frequently caught. However, the ratio of *Hylomyscus* to *Praomys* also seems to shift in favor of *Hylomyscus* in moister forest habitats (Decher et al. 2005, Malcolm and Ray 2000). More detailed ecological studies of the impact of forest disturbance on small mammal communities has also shown that *Hylomyscus* declines in favor of *Praomys* in forest islands or forests degraded by logging or agricultural activities (Malcolm and Ray 2000). Of the rodents known to be endemic to the Upper Guinean forest zone, we verified one specimen of *Malacomys cansdalei* from Draw River, but not *Hylomyscus baeri*, which is listed as "Endangered B1ab(iii)" in the recent IUCN Red List (2004). The latter species is only known from Kade in Ghana, as well as Adiopodoumé and Blékoum in Côte d'Ivoire (Heim de Balsac and Aellen 1965, Robbins and Setzer 1979). According to Grubb et al. (1999), there is a USNM specimen from Panguma, Sierra Leone.

## Bats

Table 7.3 shows an overview of bat netting results in the three reserves comprising 82 individuals and 15 species. This includes four species of fruit bats (Pteropodidae) and 11 species of insectivorous bats from three families (Nycteridae, Hipposideridae, Vespertilionidae). Total sampling effort was 214.2 net hours with an average capture success of 0.38 bats per net hour. The fruit bats included the West African endemic *Epomops buettikoferi* and the small nectar-feeding bat *Megaloglossus woermanni*. Our use of the name *Neoromicia africanus* for *Pipistrellus nanus* is based on recent revisions (Hoofer and Van den Bussche 2003, Kearney et al. 2002, Kock 2001). Our use of *Hipposideros gigas* for *H. commersoni* follows forthcoming revisions that limit *H. commersoni* to Madagascar and split the African populations into two species (Simmons in press, J. Fahr & D. Kock unpubl. data).

Draw River. At Draw River only seven captures comprising three species were made in four nights (sampling effort: 90.5 net hours, mean capture success: 0.08 bats per net hour). This included one member of the family Vespertilionidae, the Abo bat (*Glauconycteris poensis*), caught only at this site during the RAP survey, but known from several other localities in southern Ghana.

Boi Tano. At Boi-Tano nine captures were made in three nights (sampling effort: 55.0 net hours, mean capture success: 0.16 bats per net hour) including six species. Here the only member of the family Nycteridae, *Nycteris arge*, was caught, which is already known from several other localities in southern Ghana. No vespertilionids were caught at Boi-Tano. We also recorded the microbat *Hipposideros fuliginosus* at this site, as well as in Krokosua Hills. Although known from other localities in southern Ghana, this species is listed as "Near Threatened" according to the IUCN Red List (2004).

Krokosua Hills. Whereas at Draw River and Boi-Tano most nets were set in high forest or on forest paths, several nets at Krokosua Hills were set in cocoa and banana plantations and across waterholes and a small intermittent creek. The higher capture rates and greater species richness of bats found here may be a function of bats flying faster, using these areas as travel routes or for drinking, and being less cautious in avoiding obstacles such as our nets than they would be in dense forest habitat, thus increasing our capture success. Sixty-six bat captures were made including 10 species during five nights (sampling effort: 68.7 net hours, mean capture success: 0.96 bats per net hour). Three additional species (*Hypsignathus monstrosus*, *Myonycteris torquata*, *Nycteris arge*) have been recorded from this forest reserve during earlier surveys (Table 7.3; Jeffrey 1975, Van Cakenberghe and De Vree 1985, Bergmans 1989, 1997). During our survey, one *Hipposideros cyclops* and four additional vespertilionids, *Scotophilus nux*, *S. nucella*, *Pipistrellus nanulus*, and *Neoromicia africanus*, were caught here. A sizable population of the large *Hipposideros gigas*, occurring in both dark and pale color variants, used the small creek in the ravine near the "old village" of Mmem as a flyway. Many more *H. gigas* than the 12 individuals listed in Table 7.3 avoided the nets or freed themselves by chewing on the net before we could get to them. The common occurrence of *Scotophilus nucella* at Krokosua Hills (Table 7.3, Figure 7.1) is a positive surprise as this species was fairly recently described (Robbins 1984) and so far only known from eight specimens from Oda and Nkawkaw in southern Ghana, one from southeastern Côte d'Ivoire and one from western Uganda (Robbins 1984, Koopman 1989). This species, omitted in the IUCN Action Plan by Hutson et al. (2001), is listed as "Vulnerable A2a; B2ab(iii)" in the recent IUCN Red List (2004). In Ghana, *Pipistrellus nanulus* is only known from a few specimens (Ingoldby 1929, Aellen 1952, Decher et al. 1997, J. Fahr unpubl. data).

**Table 7.3:** Bat captures at Draw River, Boi-Tano and Krokosua Hills forest reserves, Ghana, 23 Oct. - 8 Nov. 2003 with some additional literature records

| Localities: | Draw River (DR) | | | | | Boi-Tano (BT) | | | | Krokosua Hills (KH) | | | | | | Grand Total |
|---|---|---|---|---|---|---|---|---|---|---|---|---|---|---|---|---|
| Date: | 23 Oct | 24 Oct | 25 Oct | 26 Oct | Total DR | 30 Oct | 31 Oct | 1 Nov | Total BT | 3 Nov | 4 Nov | 6 Nov | 7 Nov | 8 Nov | Total KH | |
| **Family/Species** | | | | | | | | | | | | | | | | |
| **Pteropodidae** | | | | | | | | | | | | | | | | |
| Epomops buettikoferi | | | | | | | 1 | | 1 | | | | | | 0 | 1 |
| Hypsignathus monstrosus | | | | | | | | | | | | | | | X[1,2] | X |
| Nanonycteris veldkampii | | | | | | 1 | 2 | | 3 | 6 | 1 | | | | 7 | 10 |
| Megaloglossus woermanni | | | | | | | | | | 4 | | 1 | | | 5 | 5 |
| Myonycteris torquata | | 1 | | | 1 | | | | | | | | | | X[1,2] | 1 |
| **Nycteridae** | | | | | | | | | | | | | | | | |
| Nycteris arge | | | | | | 1 | | | 1 | | | | | | X[3] | 1 |
| **Hipposideridae** | | | | | | | | | | | | | | | | |
| Hipposideros ruber | | | 2 | 2 | 4 | | | | | 3 | 2 | | 1 | | 6 | 10 |
| Hipposideros fuliginosus | | | | | | | 1 | | 1 | 7 | | | | 1 | 8 | 9 |
| Hipposideros beatus | | | | | | | 1 | | 1 | | | | | | 0 | 1 |
| Hipposideros cyclops | | | | | | | | | | 1 | | | | | 1 | 1 |
| Hipposideros gigas | | | | | | | | 2 | 2 | 4 | 4 | 3 | | 1 | 12 | 14 |
| Hipposideros sp.* | | | 1 | | 1 | | | | | | 1 | | 1 | 1 | 5 | 6 |
| **Vespertilionidae** | | | | | | | | | | | | | | | | |
| Pipistrellus nanulus | | | | | | | | | | | 1 | | | | 1 | 1 |
| Neoromicia africanus | | | | | | | | | | 1 | | | | | 1 | 1 |
| Glauconycteris poensis | | | | 1 | 1 | | | | | | | | | | | 1 |
| Scotophilus nux | | | | | | | | | | 1 | 1 | | 3 | 2 | 7 | 7 |
| Scotophilus nucella | | | | | | | | | | 2 | | 6 | 6 | 2 | 10 | 10 |
| Scotophilus sp.** | | | | | | | | | | | | | 2 | 1 | 3 | 3 |
| Total No. of Captures: | 0 | 1 | 3 | 3 | 7 | 2 | 5 | 2 | 9 | 31 | 10 | 4 | 13 | 8 | 66 | 82 |
| No. of Species: | 0 | 1 | 1 | 2 | 3 | 2 | 4 | 1 | 6 | 9 | 5 | 2 | 3 | 4 | 10(13) | 15(16) |
| Net Units (6 m): | 7 | 8 | 2 | 4 | | 4 | 6 | 4 | | 7 | 7 | 3 | 3 | 3 | | |
| Hours open: | 4.0 | 4.5 | 4.3 | 4.5 | | 4.3 | 4.0 | 3.5 | | *** | *** | 2.5 | 2.8 | 2.8 | | |
| Net hours (nh): | 28.0 | 36.0 | 8.5 | 18.0 | 90.5 | 17.0 | 24.0 | 14.0 | 55.0 | 23.3 | 21.3 | 7.5 | 8.3 | 8.5 | 68.7 | 214.2 |
| Netting success (bats/nh): | 0.00 | 0.03 | 0.35 | 0.17 | 0.08 | 0.12 | 0.21 | 0.14 | 0.16 | 1.33 | 0.47 | 0.53 | 1.58 | 0.94 | 0.96 | 0.38 |

* = identification of released specimens not certain, either *Hipposideros ruber* or *H. fuliginosus*; ** = identification of released specimens not certain, either *Scotophilus nux* or *S. nucella*;
*** = unequal opening periods for different nets; X[1] = literature record (Krokosua Hills; Jeffrey 1975); X[2] = literature record (Sefwi Asemparaye [sic]; Bergmans 1989; 1997); X[3] = literature record (Sefwi Mafra; Van Cakenberghe and De Vree 1985).

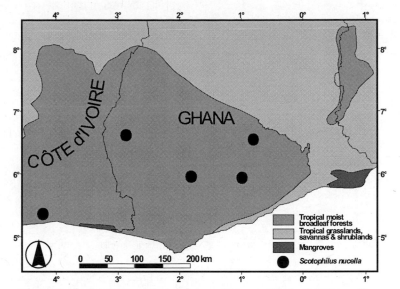

**Fig. 7.1:** Known distribution of *Scotophilus nucella* Robbins 1984 in West Africa. All records are located within the Eastern Guinean Forest ecoregion (Olson et al. 2001).

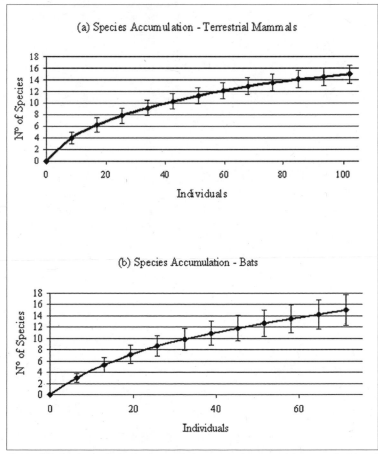

**Figure 7.2:** Randomized species accumulation curves for terrestrial mammals (a) and bats (b) sampled during the RAP survey (solid line and dots: randomization, 100 runs; vertical bars = ± 1 sd).

### Completeness of the RAP-survey

The species accumulation curves for bats rises steeply and for both terrestrial mammals and bats the curves do not reach an asymptotic plateau (Figure 7.2), indicating that our sampling of the small mammal fauna during this survey is incomplete. Based on Grubb et al. (1999), we can expect at least nine shrew and 16 small rodent species. Thus the terrestrial mammal inventory obtained 62.5% of the rodent species - 81.3% if we include the additional three species already recorded by Jeffrey (1977) for Krokosua Hills (*Graphiurus lorraineus, Mus musculoides, Lemniscomys striatus*) - and 66.7% of the shrew species expected in the forests of southwestern Ghana. However, these are conservative numbers since our survey already found two shrew species (*Crocidura foxi* and *C. obscurior*) not listed by Grubb et al. (1999). Extrapolations by one of us (JF), based on bat records from other parts of southwestern Ghana (Oda, Kade), lead to an expected 35-40 bat species in the surveyed forest reserves. This figure is 2-2.5 times higher than the currently known total of 16 bat species, emphasizing the need to complement the present inventory during subsequent surveys.

## CONSERVATION RECOMMENDATIONS

All three forests are affected by previous logging, encroaching farms, and uncontrolled hunting. At Boi-Tano one *Malacomys edwardsi* was found dead in a hunter's snare, demonstrating that even relatively small mammals are affected by hunting. Dozens of illegal snares can be found along certain trails. At Krokosua Hills locally made snap traps were used on the perimeter of a recently cleared farm surrounded by forest, presumably to keep forest rodents from raiding crops.

We recommend stricter enforcement of restrictions for hunting, logging, and the clearing of forest for cocoa plantations by the Forest Services and Wildlife Division staff, particularly in the areas designated as Globally Significant Biodiversity Areas (GSBAs). For larger mammals, but increasingly also for smaller species, there needs to be a total ban on hunting in more forests in Ghana to create refuge and recovery areas for certain species, from whence other areas can be re-populated.

Especially for bats, our short-term study may be misrepresenting true diversity in forest versus man-made habitats. A recent more long-term (one-year) study of bat communities as indicators of rainforest disturbance in Chiapas, Mexico, showed that bat capture rates were 28-64% higher in cocoa plantation, cornfield and oldfield habitat compared to forest, yet the forest site yielded 27 species, versus 21, 17, and 19 species, respectively, in the other habitats (Medellín et al. 2000).

Increasingly Ghana will also have to give thought to maintain and restore diversity of mammals in fragmented agricultural landscapes, as has been recently suggested for human-dominated landscapes in the Neotropics (Daily et al. 2003). However, this can only complement, but cannot replace, the function of protected forest areas in conserving species diversity and thus the integrity of natural ecosystems.

Villages permitted in the forest for historical or political reasons, such as the village of Mmem at Krokosua Hills, need to strictly adhere to the original village expansion limits determined around them. Villages with this exceptional status inside protected forests should receive support for developing alternative economic opportunities such as sustainable harvesting and processing of non-timber forest products (NTFPs), employment in local wildlife and forestry law enforcement, and the potentially profitable development of facilities for ecotourism (guest houses, research and visitors' centers, local guides, etc.). One example from Uganda's Bwindi Impenetrable Forest shows that involving local communities as participants in park management and providing tangible benefits through various projects, such as collection of plant resources in the forest, beekeeping societies and tree nurseries, can alleviate restrictions imposed and local resentment to conservation measures (Hamilton et al. 2000). We also feel that having no school in this village and instead sending children to boarding schools in other towns (for those who can afford it) only further alienates the young generation from understanding and living in a sustainable relationship with their forest surroundings. A recent review of the interrelationship between efforts to eliminate poverty and preserving biodiversity reaffirmed the need for reconciliation between these sometimes competing agendas, yet stressed the need for distinct approaches without compromising the retention of all living diversity (Adams et al. 2004).

In none of the schools in the villages adjacent to the sampled forests did we notice any educational posters or other learning materials such as those about rare and endangered species posted at Conservation International Ghana, or at the Wildlife Division headquarters in Accra. Such materials, including illustrated field guides (for example Kingdon 1997, van Perlo 2002), need to be made widely available in schools so that local people can gain understanding of biodiversity and conservation and gradually learn that small mammals are more than "bushmeat" or "rats" but play important ecological roles, for example as seed dispersers or predators and prey of other animals. The Wildlife Clubs of the Ghana Wildlife Society are a first step in this direction (Ntiamoa-Baidu 1995), but biodiversity education has to become a mandatory part of Ghana's educational system, as it is the key to changing currently predominant views of wildlife and forest habitat as an "inexhaustible" resource.

## REFERENCES

Adams, W.M., R. Aveling, D. Brockington, B. Dickson, J. Elliott, J. Hutton, D. Roe, B. Vira and W. Wolmer. 2004. Biodiversity conservation and the eradication of poverty. Science 306 (5699): 1146-1149.

Aellen, V. 1952. Contribution a l'étude des chiroptères du Cameroun. Mémoires de la Société Neuchâteloise Scientifique Naturelles 8: 1-121.

Animal Care and Use Committee. 1998. Guidelines for the capture, handling, and care of mammals as approved by The American Society of Mammalogists. Journal of Mammalogy 79: 1416-1431.

Bakarr, M., B. Bailey, D. Byler, R. Ham, S. Olivieri and M. Omland (eds.). 2001. From the forest to the sea: Biodiversity connections from Guinea to Togo. Conservation Priority-Setting Workshop, December 1999. Conservation International, Washington, DC, 78 pp. www.biodiversityscience.org/priority_outcomes/west_africa.

Bellier, L. 1965. Évolution du peuplement des rongeurs dans les plantations industrielles de palmier à huile. Oléagineux 20: 573-576.

Bergmans, W. 1989. Taxonomy and biogeography of African fruit bats (Mammalia, Megachiroptera). 2. The genera *Micropteropus* MATSCHIE, 1899, *Epomops* GRAY, 1870, *Hypsignathus* H. ALLEN, 1861, *Nanonycteris* MATSCHIE, 1899, and *Plerotes* ANDERSEN, 1910. Beaufortia 39: 89-153.

Bergmans, W. 1997. Taxonomy and biogeography of African fruit bats (Mammalia, Megachiroptera). 5. The genera *Lissonycteris* ANDERSEN, 1912, *Myonycteris* MATSCHIE, 1899 and *Megaloglossus* PAGENSTECHER, 1885; general remarks and conclusions; annex: Key to all species. Beaufortia 47: 11-90.

Cole, L.R. 1975. Foods and foraging places of rats (Rodentia: Muridae) in the lowland evergreen forest of Ghana. Journal of Zoology (London) 175: 453-471.

Colwell, R.K. 2004. EstimateS: Statistical estimation of species richness and shared species from samples. Version 7.0. Application and user'a guide. <purl.oclc.org/estimates>.

Daily, G.C., G. Ceballos, J. Pacheco, G. Suzán and A. Sánchez-Azofeifa. 2003. Countryside biogeography of Neotropical mammals: Conservation opportunities in agricultural landscapes of Costa Rica. Conservation Biology 17: 1814-1826.

Decher, J. 2004. A rapid survey of terrestrial small mammals (shrews and rodents) of the Forêt Classée du Pic de Fon, Guinea. Pp. 78-83. *In:* McCullough, J. (ed.). A rapid biological assessment of the Forêt Classée du Pic de Fon, Simandou Range, south-eastern Republic of Guinea. RAP Bulletin of Biological Assessment 35, Conservation International, Washington, DC.

Decher, J. and M. Abedi-Lartey. 2002. Small mammal zoogeography and diversity in West African forest remnants. Final Report. Ghana Wildlife Division, National Geographic Committee for Research and Exploration, University of Vermont, 32 pp.

Decher, J., B. Kadjo, M. Abedi-Lartey, E. O. Tounkara and S. Kante. 2005. Small mammals (shrews, rodents, and bats) from the Haute Dodo and Cavally Forests, Côte d'Ivoire. *In:* Lauginie, F., G. Rondeau and L.E. Alonso (eds.). A biological assessment of two classified forests in Southwestern Côte d'Ivoire. RAP Bulletin of Biological Assessment 34. Conservation International, Washington, DC.

Decher, J., D.A. Schlitter and R. Hutterer. 1997. Noteworthy records of small mammals from Ghana with special emphasis on the Accra Plains. Annals of Carnegie Museum 66: 209-227.

Dieterlen, F. 1976. Die afrikanische Muridengattung *Lophuromys* Peters, 1874. Stuttgarter Beiträge zur Naturkunde, Serie A (Biologie) 285: 1-96.

Fahr, J. and N.M. Ebigbo. 2003. A conservation assessment of the bats of the Simandou Range, Guinea, with the first record of *Myotis welwitschii* (Gray, 1866) from West Africa. Acta Chiropterologica 5: 125-141.

Grubb, P. and W.H.F. Ansell. 1996. The name *Graphiurus hueti* De Rochebrune, 1883 and a critique of De Rochebrune's «Fauna de la Sénégambie, Mammifères.» The Nigerian Field 61: 164-171.

Grubb, P., T.S. Jones, A.G. Davies, E. Edberg, E.D. Starin and J. E. Hill. 1999 [for 1998]. Mammals of Ghana, Sierra Leone and The Gambia. The Trendrine Press, Zennor, St. Ives, Cornwall, vi + 265 pp.

Hall, J.B. and M.D. Swaine. 1981. Distribution and ecology of vascular plants in a tropical rain forest - Forest vegetation in Ghana. Dr. W. Junk Publishers, The Hague, xv + 382 pp.

Hamilton, A., A. Cunningham, D. Byarugaba and F. Kayanja. 2000. Conservation in a region of political instability: Bwindi Impenetrable Forest, Uganda. Conservation Biology 14: 1722-1725.

Hawthorne, W.D. and M. Abu-Juam. 1995. Forest Protection in Ghana with particular reference to vegetation and plant species. Forest Inventory and Management Project Planning Branch. IUCN/ODA/Forestry Department, Republic of Ghana, Gland, Switzerland, and Cambridge, UK, xvii + 203 pp.

Heim de Balsac, H. and V. Aellen. 1965. Les Muridae de basse Côte d'Ivoire. Revue suisse de Zoologie 72: 695-753.

Holbech, L.H. 1999 [for 1998]. Small mammal survey in Ankasa and Bia protected areas. Project No. 6 ACP/GH 045. Protected Areas Development Programme, Western Region, Ghana. ULG Consultants Ltd. 67 pp.

Hoofer, S.R. and R.A. Van den Bussche. 2003. Molecular phylogenetics of the chiropteran family Vespertilionidae. Acta Chiropterologica 5(Suppl.): 1-63.

Hutson, A.M., S.P. Mickleburgh and P.A. Racey (comp.). 2001. Microchiropteran bats: Global status survey and conservation action plan, (IUCN/SSC Chiroptera Specialist Group). IUCN, Gland, Switzerland. x + 258 pp.

Hutterer, R. 1993. Order Insectivora. Pp. 69-130. *In:* Wilson, D.E. and D.M. Reeder (eds.). Mammal species of the world: A taxonomic and geographic reference.

A biological assessment of the terrestrial ecosystems of the Draw River, Boi-Tano,
Tano Nimiri and Krokosua Hills forest reserves, southwestern Ghana

65

2nd edition. Smithsonian Institution Press, Washington, DC. 1206 pp.

Hutterer, R. and D.C.D. Happold. 1983. The shrews of Nigeria. Bonner zoologische Monographien 18: 1-79.

Ingoldby, C.M. 1929. On the mammals of the Gold Coast. Annals and Magazine of Natural History (10) 3: 511-529.

IUCN. 2004. 2004 IUCN Red List of Threatened Species. <www.redlist.org>, downloaded January 2005.

Jeffrey, S.M. 1973. Notes on the rats and mice of the dry high forest of Ghana. The Nigerian Field 30: 127-137.

Jeffrey, S.M. 1975. Notes on the mammals from the high forest of Western Ghana (excluding Insectivora). Bulletin de l'Institut Fondamental d'Afrique Noire (Ser. A Sci. Nat.) 37: 950-973.

Jeffrey, S.M. 1977. Rodent ecology and land use in Western Ghana. Journal of Applied Ecology 14: 741-755.

Kearney, T.C., M. Volleth, G. Contrafatto and P.J. Taylor. 2002. Systematic implications of chromosome GTG-band and bacula morphology for Southern African *Eptesicus* and *Pipistrellus* and several other species of Vespertilioninae (Chiroptera: Vespertilionidae). Acta Chiropterologica 4: 55-76.

Kingdon, J. 1997. The Kingdon field guide to African mammals. Academic Press, San Diego, London, Boston, xviii + 464 pp.

Kock, D. 2001. Identity of the African *Vespertilio hesperida* TEMMINCK 1840 (Mammalia, Chiroptera, Vespertilionidae). Senckenbergiana biologica 81: 277-283.

Koopman, K.F. 1989. Systematic notes on Liberian bats. American Museum Novitates (2946): 1-11.

Malcolm, J.R. and J.C. Ray. 2000. Influence of timber extraction routes on Central African small-mammal communities, forest structure, and tree diversity. Conservation Biology 14: 1623-1638.

Martin, R.E., R.H. Pine and A.F. DeBlase. 2001. A manual of mammalogy: with keys to families of the world. 3rd ed. Wm. C. Brown Co., Dubuque, Iowa, xii + 436 pp.

Medellín, R.A., M. Equihua and M.A. Amin. 2000. Bat diversity and abundance as indicators of disturbance in Neotropical rainforests. Conservation Biology 14: 1666-1675.

Meester, J. and H.W. Setzer (eds.). 1971. The mammals of Africa. An identification manual. Smithsonian Institution, Washington DC. vii + 15 parts pp.

Miller jr., G.S. 1900. A collection of small mammals from Mount Coffee, Liberia. Proceedings of the Washington Academy of Sciences 2: 631-649.

Ntiamoa-Baidu, Y. 1995. Conservation education in threatened species management in Africa. Bird Conservation International 5: 455-462.

Olson, D.M., E. Dinerstein, E.D. Wikramanayake, N.D. Burgess, G.V.N. Powell, E.C. Underwood, J.A. D'Amico, H.E. Strand, J.C. Morrison, C.J. Loucks, T.F. Allnutt, J.F. Lamoreux, T.H. Ricketts, I. Itoua, W.W. Wettengel, Y. Kura, P. Hedao and K. Kassem. 2001. Terrestrial ecoregions of the world: A new map of life on Earth. Bioscience 51: 933-938. <www.worldwildlife. org/ecoregions>.

Robbins, C.B. 1984 [for 1983]. A new high forest species in the African bat genus *Scotophilus* (Vespertilionidae). Annales du Musee Royale de l'Afrique centrale, Sciences Zoologiques 237: 19-24.

Robbins, C.B. and H.W. Setzer. 1979. Additional records of *Hylomyscus baeri* Heim de Balsac and Aellen (Rodentia: Muridae) from western Africa. Journal of Mammalogy 60: 649-650.

Schlitter, D.A., L.W. Robbins and S.L. Williams. 1985. Taxonomic status of dormice (genus *Graphiurus*) from West and Central Africa. Annals of the Carnegie Museum of Natural History 54: 1-9.

Simmons, N.B. (in press). Order Chiroptera. *In:* Wilson, D.E. and D.M. Reeder (eds.). Mammal species of the world. A taxonomic and geographic reference. 3rd edition. Johns Hopkins University Press, Baltimore, MD.

Van Cakenberghe, V. and F. De Vree. 1985. Systematics of African *Nycteris* (Mammalia: Chiroptera). *In:* Schuchmann, K.-L. (ed.). Proc. Intern. Symp. African Vertebr., 53-90. Zool. Forschungsinst. Museum A. Koenig, Bonn. 585 pp.

van Perlo, B. 2003. Birds of Western and Central Africa. Princeton Illustrated Checklists. Princeton University Press, New Jersey. 384 pp.

Voss, R.S. and L.H. Emmons. 1996. Mammalian diversity in Neotropical lowland rainforests: A preliminary assessment. Bulletin of the American Museum of Natural History (230): 1-115.

Wilson, D.E., F.R. Cole, J.D. Nichols, R. Rudran and M.S. Foster (eds.). 1996. Measuring and monitoring biological diversity: Standard methods for mammals. Smithsonian Institution Press, Washington, DC. xvii + 409 pp.

# Chapter 8

## Rapid Assessment of Large Mammals at Draw River, Boi-Tano and Krokosua Hills forest reserves

*Abdulai Barrie and Oscar I. Aalangdong*

## SUMMARY

This paper presents the results of a Rapid Assessment Program survey conducted in three forest reserves in Southwestern Ghana from 22 October to 10 November 2003 to assess the biological diversity of large mammals in these areas. We used tracks, sound and visual observations, and camera phototraps to survey for the presence of large mammals. We confirmed the presence of 19, 14, and 14 species of large mammals in Draw River, Boi-Tano/Tano Nimiri, and Krokosua Hills forest reserves, respectively. In total, we confirmed the presence of 20 species of mammals in these forests. All sites are overexploited for bushmeat, encroached by farmers, and have been subject to logging. Although hunting of some species is permitted during the open season, our survey took place during the closed season (1 August – 1 December) and we found evidence for active hunting in all the forest reserves. We also found smoked duikers and monkeys carried by women bound for markets in bigger towns. Large mammals such as primates and duikers were only rarely directly observed.

## INTRODUCTION

Presently within Ghana, and elsewhere in the countries of Upper Guinea, the need to obtain information on species diversity, relative abundance and density is critical to the formulation of informed conservation and management plans. Scientists, conservationists, and policy makers want information that is reliable, accurate and as comparable as possible. Often such information is unknown, incomplete, or unavailable.

The Guinean Forest Hotspot has the highest diversity of mammals of any Hotspot with an estimated 551 species known to occur (Myers 1998, Bakarr et al. 2001). Although the number of endemic species is low, the forest is still important to the global conservation of mammals (Sayer et al. 1992, Kingdon 1997, Mittermeier et al. 1999, Davies and Hoffmann 2002) and is one of the two highest priority regions in the world for primate conservation.

The areas surveyed for this study are located within the "Guinean Forest Hotspot" as designated by CI, including the Globally Significant Biodiversity Areas (GSBAs) designated by the Government of Ghana. GSBAs are areas that have been found to harbor a high diversity of species, especially a number of globally threatened plants that are found only in Ghana. The RAP survey assessed the biodiversity of the GSBA portions of the Draw River, Boi-Tano/Tano Nimiri and Krokosua Hills forest reserves found in southwestern Ghana. The reserves surveyed are linked to Bia and Ankasa National Parks that contain remnants of wildlife populations typical of eastern Upper Guinea.

The current status of large mammals in southwestern Ghana is not well known as very little information on the presence and distribution of wildlife is known for this region (Bourlière 1963; Jeffrey 1975; Wilson 1993, 1994; Struhsaker and Oates 1995; Barnes et al. 1997; Grubb et al. 1998; Oates 1999; Oates et al. 2000).

Though large mammals such as Bongo (*Tragelaphus euryceros*) and endemic carnivores such as Johnston's civet (*Genetta johnstoni*) and Royal antelope (*Neotragus pygmaeus*) are known from small populations in Ghana (Bourlière 1963), Guinea, Liberia and Côte d'Ivoire, large mammals have not yet been systematically surveyed. In the order Artiodactyla, two threatened duikers in the genus *Cephalophus* (*C. jentinki*, and *C. zebra*) and the small Royal antelope, *Neotragus pygmaeus,* are endemic (Kingdon 1997) reinforcing the importance of the Upper Guinea "Hotspot".

Other important large mammals include the Leopard (*Panthera pardus*) and Elephant (*Loxodonta africana*). The forest elephants in Taï are considered to be priority baseline populations for West Africa (IUCN 1990, Barnes 1999). The species composition of these forest reserves is similar to that of Bia and Ankasa. However, no systematic surveys of large terrestrial mammals have been undertaken in these three sites.

The greater Mount Nimba, Ziama, Badiar, Taï, Gola, Tiwai, Sapo, Ankasa and Bia forest reserves, and other forest reserves in this region, contain a significant representation of biodiversity of what remains of West African rainforests. The objective of this RAP survey was to provide quick, efficient, reliable, and cost-effective biodiversity data on this little known region of southwestern Ghana as part of a regional conservation strategy.

## MATERIALS AND METHODS

### Study Area
We conducted our surveys at the end of the rainy season in three forest reserves in Southwestern Ghana:

Site 1 was the Draw River forest reserve near Ankasa (5°11'35. 8"N, 2°24'26. 2"W) from 22 – 27 October 2003;

Site 2, Boi-Tano/Tano Nimiri forest reserves (5°31'55. 1"N, 2°37'6. 5"W) from 28 October to 2 November 2003; and

Site 3, Krokosua Hills forest reserve (6°36'46. 7"N, 2°50'58. 3"W) from 3 - 9 November 2003.

### Methods
We used active and passive methods to document the presence of large mammals. The active method included direct observation of species and track and sound identification, nests, dung and other indirect information to determine presence of large mammalian species in the three study areas. Direct observations and track and sound identification were made during daily excursions from base camp. Surveys were carried out at night using a spotlight. Because our colleagues also collected our records opportunistically and some observations may have been repeated we used this information only to document species presence.

The passive method included the use of nine CamTrakker phototraps (CamTrakker Atlanta, Georgia) operated at each study site. CamTrakker phototraps are triggered by heat-in-motion. Each CamTrakker used a Samsung Vega 77i 35mm camera set on auto focus and loaded with Kodak Gold 200 print film. Time between sensor reception and a photograph was 0.6 secs. Cameras were set to operate con-

tinuously (control switch 1 on) and to wait a maximum 20 seconds between photographs (control switches 6 and 8 on). Cameras were placed at sites suspected of being frequented by various mammalian species. Den sites, trails, and feeding stations such as fruiting trees were typically chosen for camera placement. Cameras were located approximately 800 m apart and at least 1,500 m from base camp. We use this method to calculate observation rates for each site just as standard transects are used. Instead of the observer making observations along a route, "observations" moved along routes in front of fixed cameras (observers). For shy mammals under severe hunting pressure, phototrapping methods can be more effective than walking transects, especially when observers have different and varied levels of expertise.

## RESULTS

We observed, identified by sound or photographed 19, 14 and 14 species of large mammals in Draw River, Boi-Tano/Tano Nimiri and Krokosua Hills, respectively (Appendix 11) for a combined total of 20 species of large mammals. The camera phototraps obtained no photograph at Draw River or Boi-Tano/Tano Nimiri and one photograph of the Giant rat (*Cricetomys* spp.) at Krokosua Hills. No Yellow-backed duiker, Leopard, Aardvark, Pangolin or Bongo were observed but local hunters reported that these species still occur in these forest reserves. Using tracks, dung and other signs, we documented the presence of the Forest elephant (*Loxodonta africana cyclotis*) in the Draw River forest reserve. Tracks and dung of Forest elephants were common in all forested areas of the Draw River forest reserve.

Mammals documented at only one forest reserve were Red-river hog, Forest elephant, Giant forest hog, and Pardine genet (Appendix 11). We believe these differences are due to the short duration of our survey and also likely a result of fundamental differences in mammalian faunas in the study areas.

Results from interviews and direct observations in the field indicated that there is a thriving bushmeat trade in southwestern Ghana. Three smoked brush-tailed porcupines, two giant pouched rats and one tail of a Beecroft's anomalure were found in the Draw River forest reserve. Two smoked duikers and one smoked monkey were found in Boi Tano/Tano Nimiri. The fur of a Royal antelope was found at a hunting camp in the Krokosua Hills forest reserve (Appendix 11). These represent the highest number of animals seen during the entire survey.

All hunting in Ghana is restricted and requires permits during the open season. During the closed season (1 August -1 December) hunting is not permitted. Ghanaian laws also prohibit the hunting of species of global and national concern. Our survey took place during the closed season when permits are not issued for hunting and we found 125 shotgun shells and heard gunshots during our daily routine. We found seven active hunting camps. A total of 96, 43 and 59 snares were found in Draw River, Boi-Tano/Tano Nimiri and Krokosua Hills respectively. In Krokosua Hills, a tree of 40 cm dbh (diameter at breast height) was felled with an axe to capture one tree hyrax.

## DISCUSSION

Large mammals in Ghana and throughout much of West Africa are extremely rare as a result of unregulated exploitation, habitat loss and the increasing demand for bushmeat (Lowes 1970, Davies 1987, Starin 1989, Martin 1991, Fa et al. 1995, McGraw 1998). Many forest-dependent animals are now rare, and the demand for bushmeat has further reduced wild populations to such low levels that a number of them can no longer be considered viable.

Despite the existence of reports (Bourlière 1963, Jeffrey 1975, Grubb et al. 1998) indicating high diversity and density of large mammals, we were able to confirm the presence of only 20 species of large mammals occurring at relatively low density (Appendix 11). The diversity and density of large mammals we recorded in Ghana is the lowest compared to results from other RAP surveys in Guinea and Côte d'Ivoire (Bakarr and Struhsaker 1999, Barrie and Kante 2004, Sanderson and Trolle 2005). The current distribution pattern of most large mammals in the areas surveyed reflects the increased pressure on their habitats and populations.

Logging is locally intense and destructive in many countries in West Africa and has been cited as the primary cause of habitat destruction in Ghana. Global demand for valuable hardwoods continues to spur logging in what remains of Upper Guinea forests. Timber harvesting has accelerated forest fragmentation and facilitated the loss of large mammals. Moreover, loggers often support themselves and their families on bushmeat, consume trees for fuelwood, and create extensive networks in forests to exploit them. The secondary impacts of logging are equally as destructive to the forest. Logging roads create access to the interior of the forests bringing hunters into areas that were otherwise not penetrable (Oates 1999, Sayer et al. 1992, Wilkie et al. 1992). These destructive forces operating throughout greater West Africa are operating in Draw River, Boi-Tano/Tano Nimiri and Krokosua Hills forest reserves. In particular the Boi-Tano/Tano Nimiri forest reserve was still being enumerated for logging.

People living in Ghana like most other countries in West and Central Africa have traditionally hunted and relied on bushmeat to provide them with protein (Barnes and Lahm 1997, Wilkie and Carpenter 1999). The supply of bushmeat is a lucrative business in Ghana as in other parts of Africa (Asibey 1976, Jeffrey 1977, Ajayi 1979, Martin 1983, Oates 1986, Falconer and Koppell 1990, Njiforti 1996, Bowen-Jones and Pendry 1999). Bushmeat hunting parallels habitat loss as a major threat to the survival of mammals in West Africa (Bakarr and Struhsaker 1999, Bakarr et al. 2001). In West Africa, duikers and primates are the species most preferred by hunters (Eves and Bakarr 2001).

Antelopes, forest pigs, and primates dominate the bushmeat trade, while cane rat (*Thryonomys swinderianus*) and giant rat (*Cricetomys* spp.) are consumed by rural people. The recent extinction of *Procolobus badius waldroni* in West Africa was attributed to hunting and the demand for bushmeat in this region (Oates et al. 2000).

Hunting is a major source of meat around the forest reserves surveyed during the RAP. While our assessment lasted less than a month, certain species were strikingly absent including *Cephalophus rufilatus*, *Cephalophus silviculture*, *Syncerus caffer*, *Tragelaphus euryceros*, and *Panthera pardus*. The amount of bushmeat coming out of the forest reserves appears to have increased dramatically in the last decade. The extent of such hunting has prompted governments to enact hunting bans, though the legislation is often impractical and cannot be enforced (Sayer et al. 1992).

During our survey we utilized an extensive network of trails created by hunters and new and old logging roads created by heavy machinery used to extract and transport cut trees from the forests. Parts of the Boi-Tano/Tano Nimiri reserves had already been enumerated for logging. These trail networks fragment forest reserves, reducing the area for wildlife. Collateral damage from logged trees was extensive and extraction procedures had damaged many untargeted trees in the reserves surveyed.

Local communities continue to impact forest reserves by hunting and establishing farms within reserve boundaries, leading to additional habitat degradation. Local pressures for fuelwood (e.g. charcoal), bushmeat, farm- and cropland, and global demand for timber and mineral resources such as gold and bauxite, are reducing the size and future potential of remaining forests throughout West Africa. Primary forests outside protected areas are targeted for timber extraction and secondary forests are being encroached upon. Much of the forest in this region has undergone vast changes in area and composition as a result of habitat fragmentation. In addition, the increase in human population is accelerating the conversion of remaining forest habitats into human-dominated settlements and agricultural landscapes.

The combined forces of indiscriminate logging practices, illegal hunting, and human encroachment into logged forests act strongly against the perpetuation of large mammals in Draw River, Boi-Tano/Tano Nimiri and Krokosua Hills forest reserves. Our results are suggestive of and consistent with the "empty forest syndrome" whereby large mammal populations are one by one reduced in density and finally extirpated from forested areas. Unless logging practices are reduced and controlled, poaching is stopped altogether, and encroachment ceases, the long-term outlook for large mammals in Draw River, Boi-Tano/Tano Nimiri and Krokosua Hills forest reserves is most certainly grim.

Nevertheless, our results suggest the full biologically rich assortment of large mammals present in Ghana's National Parks are also found in the Draw River, Boi-Tano/Tano Nimiri and Krokosua Hills forest reserves. During our brief visit we documented the presence of the threatened Forest elephant and Royal antelope, suggesting that the inclusion of Draw River, Boi-Tano/Tano Nimiri and Krokosua Hills forests into a protected area system can act to conserve and increase the regional populations of these species.

## CONSERVATION RECOMMENDATIONS

The future of large mammals in Ghana is bleak. The most serious threat to the future survival of large mammals in Ghana is the destruction of their habitat. Wildlife regulations must be strenuously enforced in these areas for species to be conserved. There is also the need for extensive surveys to establish a complete list of large mammal species in these areas. The local people must also receive education on wildlife laws. Conservation action needs to be designed to protect what the laws state must be protected. Recommendations are of no value unless they can be initiated under the umbrella of enforcement. Unless timely and consistent progress is made by government agencies, conservation organizations and other stakeholders in Ghana, large mammal conservation cannot succeed.

Unless logging activities are carefully regulated, bushmeat hunting is halted and environmental awareness is intensified, Draw River, Boi-Tano/Tano Nimiri and Krokosua Hills forest reserves will soon become open woodlands that support none of the rich biodiversity found in primary forests. Because large mammals in the reserves continue to be threatened even in the wake of the current efforts by the Forestry Commission, urgent efforts should be made to ensure their long-term survival. It is in light of this that the following recommendations are being made:

a) *Habitat and Wildlife Protection.* Despite their status as GSBAs, the Draw River, Boi-Tano/Tano Nimiri and Krokosua Hills forest reserves continue to experience uncontrolled exploitation of both wildlife and plant resources. Government agencies with responsibilities for controlling illegal exploitation in the reserve lack the resources, the capacity, and, traditionally, the mandate, to monitor and enforce laws, especially with regard to wildlife. Ensuring that the habitat and the wildlife are protected from destructive activities requires training and equipping wildlife and forestry personnel to control illegal activities. A possible solution to the lack of sufficiently large areas of suitable habitat would be the re-designation of production forest to wildlife reserves, if sufficient pressure could be exerted. It is particularly important to include Draw River as part of Ankasa Resource Reserve and Nini Suhien National Park. This will provide a larger and much needed habitat for the conservation of the globally threatened Forest elephant, *Loxodonta africana*.

b) *Community Involvement.* Because government agencies with responsibility for managing the reserves cannot do it alone, there is a need to involve local communities in the management of the reserves as the destruction of local resources will greatly affect their well-being. Co-management arrangements involving well-defined responsibilities for all parties (communities and government) will provide opportunities for involvement in controlling and regulating access to the reserves and resources.

c) *Scientific Research.* While there has been ongoing research on forest biodiversity throughout Ghana, information about the wildlife and plants of the forest reserves is still scanty. There is a need to establish a scientific presence in these reserves to generate the relevant data for appropriate management. In addition, the presence of a scientific team on-site can serve to ward off would-be trespassers. In this regard, it is important that conservation NGOs collaborate with universities and research institutions. Monitoring large mammals using reliable methodologies such as photo-trapping (Karanth and Nicholas 1998, Linkie et al. 2003) and transect sampling can help to determine population trends and to allow for the evaluation of management strategies.

d) *Hunting and Trapping of Bushmeat.* A number of species, especially larger mammals, are threatened as a result of the high demand for bushmeat in both urban and rural markets. Providing alternative sources of protein will ensure that most of the threatened wildlife resources may continue to survive. Moreover, the extensive use of snares and other trapping methods should be addressed by enforcing current wildlife laws. Large mammal populations are declining in part as a result of excessive trapping with snares.

e) *Logging.* Logging activities have in the past and will potentially continue to impact the Boi-Tano/Tano Nimiri forest reserves of which a central part has already been delineated for logging. The logging of this segment will essentially fragment the reserve into two halves. Meanwhile, logging companies appear not to be involved in any activities to control the illegal exploitation of resources in the reserve, especially in areas bordering their jurisdiction. More often than not, company employees have been observed to be involved in the illegal exploitation of bushmeat and also provide a ready market for bushmeat hunters. Habitat destruction through logging is the most serious threat to large mammals. Logging should be halted and the enumeration currently underway in the Boi-Tano/Tano Nimiri stopped to allow for the restoration of the population of large mammals. In most cases, this activity provides the opportunity for hunters to gain access to otherwise inaccessible areas of the forest reserve. Initiating a reforestation program that utilizes indigenous timber species could rehabilitate some degraded areas of the reserve.

f) *Demarcate Reserve Boundary.* The boundaries of these reserves are currently unknown even to the wildlife guards and rangers responsible for controlling illegal activities. Every effort should be made to ensure that the reserve boundaries are clearly demarcated to prevent illegal encroachment by local communities, especially in the form of coffee and cocoa plantations.

# REFERENCES

Ajayi, S.S. 1979. Food and animal production from tropical forest: utilization of wildlife and by-products in West Africa. FAO, Rome, Italy.

Asibey, E.O.A. 1974. Wildlife as a source of protein in Africa south of the Sahara. Biological Conservation 6(1): 32-39.

Asibey, E.O.A. 1976. The effects of land use patterns on future supply of bushmeat in Africa south of the Sahara. Working Paper on Wildlife Management and National Parks, 5th Session.

Bakarr, M.I. and T.T. Struhsaker. 1999. A Rapid Survey of Primates and other Mammals in Parc National de la Marahoué, Côte D'Ivoire. In: Schulenberg, T.S., C.A. Short and P.J. Stephenson (eds.) A Biological Assessment of Parc National de la Marahoué. RAP Working Papers 13. Conservation International, Washington, DC.

Bakarr, M.I., G.A.B. da Fonseca, R. Mittermeier, A.B. Rylands and K.W. Painemilla (eds.). 2001. Hunting and Bushmeat Utilization in the African Rain Forest. Advances in Applied Biodiversity Science. 2: 5-170.

Barnes, R.F.W., B. Asamoah-Boateng, J. Naadja Majam and J. Agyei-Ohemeng. 1997. Rainfall and the population dynamics of elephant dungpiles in the forests of southern Ghana. African Journal of Ecology, 35:39-5239-52.

Barnes, R.F.W. 1999. Is there a future for elephants in West Africa? Mammal Review, 29:175-199.

Barnes, R.F.W. and S.A. Lahm. 1997. An ecological perspectives on human densities in the central African forests. Journal of Applied Ecology, 34:245-260.

Barrie, A. and S. Kante. 2004. A rapid survey of the large mammals of the Forêt Classé du Pic de Fon, Guinea. In: McCullough, J. (ed.). A Rapid Assessment of the Forêt Classée du Pic de Fon, Simandou Range, South-eastern Republic of Guinea. RAP Bulletin of Biological Assessment 35, Conservation International, Washington, DC.

Bourlière, F. 1963. Observations on the ecology of some large African mammals. In: Howell, F.C. and F. Bourlière (eds.). African ecology and human evolution. Aldine Pub. Co., Chicago. Pp. 43-54.

Bowen-Jones, E. and S. Pendry. 1999. The threat to primates and other mammals from the bushmeat trade in Africa, and how this threat could be diminished. Oryx 33(3):233-246.

Davies, G. and M. Hoffmann. (eds). 2002. African Forest Biodiversity. A Field Survey Manual for Vertebrates. Earthwatch Europe.

Davies, A.G. 1987. Conservation of primates in the Gola Forest reserves, Sierra Leone. Primate Conservation, 8:151-153.

Eves, H.E. and M.I. Bakarr. 2001. Impacts of bushmeat hunting on wildlife populations in West Africa's Upper Guinea Forest Ecosystem. In: Bakarr, M.I., G.A.B. da Fonseca, R. Mittermeier, A.B. Rylands and K.W. Painemilla (eds.). Hunting and Bushmeat Utilization in the African Rain Forest: Perspectives Toward a Blueprint for Conservation Action.. Conservation International, Washington, DC. Advances in Applied Biodiversity Science Number 2:39-57.

Fa, J. E., J. Juste, J. Perez del Val and J. Castroviejo. 1995. Impact of market hunting on mammal species in Equatorial Guinea. Conservation Biology 9:1107-1115.

Falconer, J. and C. Koppel. 1990. The Major Significance of Minor Forest Products: The Local Use and Value of Forests in the West African Humid Forest Zone. FAO, Community Forests Note 6. Rome.

Grubb, P., T.S. Jones, A.G. Davies, E. Edberg, E.D. Starin and J.E. Hill. 1998. Mammals of Ghana, Sierra Leone and The Gambia. The Tendrine Press, Zennor, St Ives, Cornwall, UK. 265 p.

IUCN. 1990. 1990 IUCN Red List of Threatened Animals. IUCN, Gland Switzerland and Cambridge, UK. 228 pp.

Jeffrey, S.M. 1975. Notes on the mammals from the high forest of Western Ghana (excluding Insectivora). Bulletin de l'Institut Fondamental d'Afrique Noire, 37:950-973.

Jeffrey, S. 1977. How Liberia uses wildlife. Oryx 14:168-173.

Karanth, K.S. and J.D. Nicholas. 1998. Estimation of tiger densities in India using photographic captures and recaptures. Ecology, 79, 2852-2862.

Kingdon, J. 1997. The Kingdon Field Guide to African Mammals. Harcourt Brace & Company, New York.

Linkie, M., D.J. Martyr, J. Holden, A. Yanuar, A.T. Hartarna, J. Sugardjito and N. Leader-Williams. 2003. Habitat destruction and poaching threaten the Sumatran tiger in Kernci Seblat National Park, Sumatra. Oryx, 37 (1) 41-48.

Lowes, R.H.G. 1970. Destruction in Sierra Leone. Oryx 10(5): 309-310.

Martin, L.G.H. 1983. Bushmeat in Nigeria as a natural resource with environmental implications. Environmental Conservation 10(2): 125-132.

Martin, C. 1991. The rainforests of West Africa: Ecology, Threats and Conservation. Birhauser Verlag, Boston.

McGraw, W.S. 1998. Three Monkeys nearing extinction in the forest reserves of eastern Côte d'Ivoire. Oryx 32(3):233-236.

Mittermeier. R.A., N. Myers and C.G. Mittermeier (eds.). 1999. Hotspots: earth's biologically richest and most endangered terrestrial ecoregions. Cemex, Conservation International. 430 p.

Myers, N. 1998. Threatened biotas: 'hotspots' in tropical forests. Environmentalist 8:187-208.

Njiforti, H.L. 1996. Preferences and present demand for bushmeat in north Cameroon: some implications for wildlife conservation. Environmental Conservation 23(2):149-155.

Oates, J.F. 1986. Action plan for African primate conservation 1986-1990. IUCN/SSC. Primate Specialist Group, New York, USA.

Oates, J.F. 1999. Myth and reality in the rainforest: how conservation strategies are failing in West Africa. University of California Press, Berkeley, xxviii+310 pp.

Oates, J.F., M. Abedi-Lartey, S. McGraw, T.T. Struhsacker and G.H. Whitesides. 2000. Extinction of a West African red colobus monkey. Conservation Biology, 14:1526-1532.

Sanderson, J.G. and M. Trolle. 2005. Monitoring elusive mammals. American Scientist 93(2): 148-155. DOI: 10.1511/2005.2.148.

Sayer, J.A., C.S. Harcourt and N.M. Collins (eds.). 1992. The Conservation Atlas of Tropical Forests; Africa. IUCN, Simon and Schuster, New York. 288 p.

Starin, E.D. 1989. Threats to the monkeys of The Gambia. Oryx 23:385-391.

Struhsaker, T.T. and J.F. Oates. 1995. The biodiversity crisis in southwestern Ghana. Primates 1, 5-6.

Wilkie, D.S. and J.F. Carpenter. 1999. Bushmeat hunting in the Congo Basin: An assessment of impacts and options for mitigation. Biodiversity and Conservation, 8:927-955.

Wilkie, D.S., J.G. Sidle and G.C. Boundzanga. 1992. Mechanised logging, market hunting, and a bank loan in Congo. Conservation Biology 6(4): 570-580.

Wilson, D.E. 1993. Mammal species of the world: A taxonomic and geographic reference. 2nd edition. Smithsonian Institution Press, Washington, DC. 1206 pp.

Wilson, J.V. 1994. Final Report. Three-year survey of the duikers of Ghana (1991-1993). Chipangali Wildlife Trust (Zimbabwe) and Game and Wildlife Dept./IUCN Project 9786 Accra-Ghana.

# Chapter 9

## Rapid Assessment of the Primates of Draw River, Boi-Tano and Krokosua Hills

*Tobias Deschner and David Guba Kpelle*

## SUMMARY

We conducted a rapid assessment of the primate fauna between October 23 and November 9, 2003 in four forest reserves in southwestern Ghana: Draw River, Boi-Tano, Tano Nimiri and Krokosua Hills forest reserves. We used line transect methods, "scouting surveys" and interviews with hunters to estimate the presence and abundance of primate species. In total we confirmed the presence of six different species, including two prosimians (*Perodicticus potto potto* and *Galagoides demidovii*), three anthropoid monkeys (*Cercopithecus petaurista, Cercopithecus lowei* and *Procolobus verus*) and one hominoid ape, the West African chimpanzee (*Pan troglodytes verus*). Local hunters indicated the presence of three other anthropoid monkey species *(Cercopithecus diana roloway, Cercocebus atys lunulatus* and *Colobus vellerosus*). While the presence of these species is very likely and is confirmed by the results of other surveys, that of Miss Waldron's Red Colobus (*Procolobus badius waldroni*), although reported by hunters in three different reserves, remains highly speculative. Sightings of all diurnal primate species were too rare to allow for reliable density estimates. This prevalent situation is largely attributable to intense hunting pressure and habitat destruction. Taking into consideration the general precarious status of threatened primate species in Ghanaian forest reserves, we recommend increased and rapid protection measures in the surveyed forest blocks to prevent the local extinction of threatened primate species.

## INTRODUCTION

Primate species represent an important part of tropical ecosystems. In some tropical forests, primates make up to 46% of the total mammalian biomass (Terborgh 1983). They occupy an important role in pollination, seed dispersal and seed predation. In fact, due to their numbers frugivorous primates might be the most important seed dispersers in tropical forests (Chapman 1995, Chapman and Onderdonk 1998).

Primates also form a crucial part in the tropical forest food web, as prey for different predator species such as raptors, snakes and felids (Cheney and Seyfarth 1990, Struhsaker and Leakey 1990, Boesch 1991, Cowlishaw 1994) as well as predators of species such as insects, squirrels, duikers, and other primates (Hausfater 1976, Bearder 1987, Boesch and Boesch 1989). Due to their relatively large body size and conspicuous diurnal lifestyle, a vast number of primate species figure under the preferred game for hunters (Caspary et al. 2001). Together with habitat destruction, hunting is the major threat for survival for primates in tropical ecosystems (Mittermeier and Cheney 1987, Cowlishaw and Dunbar 2000).

These threats have led to a rapid decline in wild primate population size in a vast number of species. For certain species, like chimpanzees, population numbers have declined by 66% within the last three decades (Butynski 2001).

The forest reserves surveyed during this RAP, Draw River, Boi-Tano, Tano Nimiri and Krokosua Hills, are all located in the southwestern part of Ghana. This region is part of the Upper Guinea forest ecosystem, which includes forests from eastern Sierra Leone to eastern Togo and is considered one of the world's 25 priority conservation areas because of its high degree of biodiversity and endemism (Mittermeier et al. 1998).

Southwestern Ghana is characterized by wet evergreen to moist semidecidous forests. In these forests, the following primate species can be expected: Two prosimian species, Demidoff's Dwarf Galago (*Galagoides demidovii*) and the Potto (*Perodicticus potto potto*), three guenon species, the Roloway Monkey (*Cercopithecus diana roloway*), Lowe's Monkey (*Cercopithecus campbelli lowei*) and the Lesser Spot-nosed Monkey (*Cercopithecus petaurista*), three Colobus species, the White-thighed Colobus (*Colobus vellerosus*), Miss Waldron's Red Colobus (*Procolobus badius waldroni*) and the Olive Colobus (*Procolobus verus*), the White-naped Mangabey (*Cercocebus atys lunulatus*) and the Western Chimpanzee (*Pan troglodytes verus*).

However numerous reports conducted in a wide range of different forest patches in Southwestern Ghana during the last decades have reported declining numbers of primate populations or have even failed to confirm the presence of certain monkey species (for a summary see Oates et al. 1996 and Oates et al. 2000). In fact, the first species considered to become extinct within the last century is Miss Waldron's Red Colobus that once inhabited the rain forests of eastern Côte d'Ivoire and southwestern Ghana (Oates et al. 2000). These reports underline the fact that primate populations in these areas are steadily declining and face a serious extinction risk if protection status of the remaining habitats is not improved.

This survey was conducted to assess the presence and relative abundance of primate populations in four different forest blocks in southwestern Ghana to estimate the threats primates face in these reserves and to propose recommendations for the necessary protection measures.

## METHODS

We conducted a census at four different forest reserves, Draw River, Boi-Tano, Tano Nimiri and Krokosua Hills in the southwestern part of Ghana between October 23 and November 9, 2003.

Surveys were conducted by the two authors and a local guide by walking slowly (300-1,500 m/hr) and quietly, watching out for any indication of primate presence. Transects either followed along poaching trails, logging roads, logging compartment lines or straight compass bearing.

Specifically we looked for movements in trees, chimpanzee nests, feeding remains on the ground and we listened for primate vocalizations and movements in trees. When encountering primates we noted date, time of day, location and transect number, species detected (estimated number of individuals when possible), method of detection (visual or auditory), behavior and distance from observer and angle from transect line. The low number of direct and indirect signs of primates prevented us from calculating density estimates for the different species.

In addition to transects, we used "Scouting surveys" to detect primate presence by following existing trails and stopping regularly to listen for primate vocalizations. When detecting primates, we approached quietly to determine the species and numbers.

We spent a total of 134 h and 29 min on transects and scouting surveys. At Draw River, we conducted 4 transects lasting 8 h 33 min with an additional scouting survey time of 32 h 27 min. Two transects were carried out at Boi-Tano lasting 3 h 08 min and the three transects carried out at Tano Nimiri lasted 7 h 35 min. Scouting survey time for these two areas added up to 38 h 28 min. Finally, we conducted two transects at Krokosua Hills that lasted 6 h 23 min and scouting surveys of 37 h 55 min duration (Table 9.1).

The censuses were carried out between 3:30 a.m. and 20:30 p.m. On several occasions we camped overnight in remote areas of the reserves to be able to monitor nocturnal vocalizations and start censuses early the following morning.

At all sites we conducted interviews with local hunters, wildlife guards and Community Biodiversity Advisory Groups (voluntary monitoring groups for illegal activities run by local communities). We presented them with pictures of different primate species and asked about their presence in the forest as well as the last time they had seen each species. Additionally, we asked hunters about their hunting methods, age, marital status, number of children and occupation (farm type and size).

## RESULTS

Overall, we confirmed the presence of two prosimian species *Galagoides demidovii*, "Demidoff's Dwarf Galago" and *Perodicticus potto*, "Potto" for all sites and three anthropoid

**Table 9.1.** Summary of survey activities in the four forest reserves

| | Draw River | Boi-Tano | Tano Nimiri | Krokosua Hills | Total |
|---|---|---|---|---|---|
| Time spent on transects | 8h 33min | 3h 08min | 7h 35min | 6h 23min | 25h 39min |
| Sum of transect distances | 7400m | 2775m | 3351m | 3967m | 17493m |
| Time spent scouting | 32h 27min | 15h 53min | 13h 35min | 37h 55 min | 99h 50min |
| Interviews with hunter | 3 | 1 | 2 | 3 | 9 |

monkey species *Cercopithecus campbelli lowei*, "Lowe's Monkey", *Cercopithecus petaurista*, "Lesser Spot-nosed Monkey and *Procolobus verus*, "Olive Colobus" for one or more sites. The presence of one hominoid ape species, *Pan troglodytes verus*, the Western Chimpanzee, was confirmed at two sites (Table 9.2).

Interviews with local poachers indicate the presence of additional primate species (*Cercopithecus diana roloway*, "the Roloway Monkey", *Colobus vellerosus*, "White-thighed Colobus", *Cercocebus atys lunulatus*, "White-naped Mangabey; Table 9.3). At three of the sites hunters claimed to have recently seen individuals of *Procolobus badius waldroni*, Miss Waldron's Red Colobus, a species considered to be extinct (Oates et al. 2000).

The abundance of all confirmed diurnal primates was so low that no density calculations could be conducted.

At Draw River, we visually confirmed the presence of Lowe's Monkey on three occasions, of the Lesser Spot Monkey guenon on two occasions and the Olive Colobus on one occasion (Table 9.2).

At Boi-Tano and Tano Nimiri no diurnal monkey presence could be confirmed through visual or acoustical signs.

At Krokosua Hills, we visually confirmed the presence of Lowe's Monkey on one occasion and the presence of the Lesser Spot-nosed Monkey on one occasion. Long distance vocalizations of Lowe's Monkey could be heard by the two authors on two occasions and by other members of the RAP team on one occasion.

On two occasions we found feeding remains of chimpanzees (piles of *Sacoglottis gabonensis* kernels with food wedges) at Draw River. We found one nest group consisting of five chimpanzee nests at Tano Nimiri. The height of the nests (0-3 m), built in *Scaphopetalum amoenum* bushes, and the fact that four of the nests were fresh but without any

signs of urine or feces around them indicated that they were day nests. Additionally we found feeding remains consisting of *Chrysophyllum subnudum* fruits with teeth imprints of chimpanzees on one occasion at Tano Nimiri.

At Krokosua Hills we found feeding remains consisting of *Marantochlea leucantha* stems, which the local guide claimed come from chimpanzees.

At all four sites we found signs of human presence and heavy poaching activity. At Draw River we found a dense network of poaching trails combined with a large number of snares. Cartridges were found regularly, however gun shots were heard infrequently indicating a very low density of adequate game. On the first day at Draw River we found a poaching camp with two hunters in it that was subsequently destroyed by guards of the Forest Services Division. The density of poaching trails seemed to be lower at Boi-Tano and Tano Nimiri and the number of encountered snares was lower as well. However, the high number of cartridges found in the two forest blocks indicated a high poaching activity, which was confirmed by local hunters. In addition to intense poaching, the Krokosua Hills reserve suffers from a high pressure of illegal farming and logging activities. Illegal loggers reportedly have attacked guards of the forestry department and our local guides refused to approach the noise of chain saws when we heard them close by in the reserve. Due to these activities, the forest at Krokosua Hills appeared to be the most disturbed of all the forest reserves visited.

We listed fruit species found in the different forest reserves that are known to be consumed by chimpanzees in Taï National Park in Côte d'Ivoire. At Draw River we observed the presence of the following fruit: *Parinari excelsa, Sacoglottis gabonensis, Landolphia hirsuta, Dialium aubrevillei, Panda oleosa, Irvingia gabonensis,* and *Strychnos aculeata.* At Boi-Tano and Tano Nimiri we found fruits of

**Table 9.2.** Presence of primate species in the four forest reserves recorded during survey walks. Explanation of symbols: + = species present; - = species absent; (+) = species encountered by other RAP members; S = sighted; H = heard; N = nest; FR = feeding remains.

| Species | Vernacular name | Draw River | Boi-Tano | Tano Nimiri | Krokosua Hills | Confirm. |
|---|---|---|---|---|---|---|
| *Perodicticus potto potto* | Potto | (+) | - | - | - | S(1) |
| *Galagoides demidovii* | Demidoff's Dwarf Galago | + | + | + | + | S(1), H(X) |
| *Cercocebus atys lunulatus* | White-naped Mangabey | - | - | - | - | - |
| *Cercopithecus diana roloway* | Roloway Monkey | - | - | - | - | - |
| *Cercopithecus campbelli lowei* | Lowe's Monkey | + | - | - | + | S(X), H(X) |
| *Cercopithecus petaurista petaurista* | Lesser Spot-nosed Monkey | + | - | - | + | S(X), H(X) |
| *Colobus vellerosus* | White-thighed Colobus | - | - | - | - | - |
| *Procolobus badius waldroni* | Miss Waldron's Red Colobus | - | - | - | - | - |
| *Procolobus verus* | Olive Colobus | + | - | - | - | S(X) |
| *Pan troglodytes verus* | Western Chimpanzee | + | - | + | - | N(1), FR(2) |

**Table 9.3.** Likely presence of primate species in the four forest reserves indicated by hunters. Explanation of symbols: + = species present; - = species absent; "-/+" = contradicting information.

| Species | Vernacular name | Draw River | Boi-Tano | Tano Nimiri | Krokosua Hills |
|---|---|---|---|---|---|
| *Perodicticus potto potto* | Potto | + | + | + | + |
| *Galagoides demidovii* | Demidoff's Dwarf Galago | + | + | + | + |
| *Cercocebus atys lunulatus* | White-naped Mangabey | + | + | + | + |
| *Cercopithecus diana roloway* | Roloway Monkey | "-/+" | + | + | + |
| *Cercopithecus campbelli lowei* | Lowe's Monkey | + | + | + | + |
| *Cercopithecus petaurista petaurista* | Lesser Spot-nosed Monkey | + | + | + | + |
| *Colobus vellerosus* | White-thighed Colobus | "-/+" | + | + | + |
| *Procolobus badius waldroni* | Miss Waldron's Red Colobus | - | + | + | + |
| *Procolobus verus* | Olive Colobus | + | + | + | + |
| *Pan troglodytes verus* | Western Chimpanzee | + | + | + | + |

*Dialium aubrevillei, Strychnos aculeata, Panda oleosa, Parinari excelsa, Sacoglottis gabonensis, Klainedoxa gabonensis, Mammea africana, Coula edulis, Diospyros sansa-minika, Irvingia gabonensis, Chrysopyllum subnudum* and *Uapaca esculenta*. We also found leaf remains of *Parkia bicolor* apparently fed on by either chimpanzees or monkeys of the Colobus family. The most probable is the White-thighed Colobus (*Colobus vellerosus*; Korstjens, personal communication). The forest at Krokosua Hills was so different in plant composition that no comparable fruit list could be established for that site.

## DISCUSSION

Primates are long-lived, large bodied and slowly reproducing animals. These characteristics make them especially vulnerable to environmental changes such as habitat destruction and hunting pressure. Because diurnal primates live in complex structured social groups, intense hunting pressure not only decimates absolute numbers but also alters natural group composition and thereby destroys long lasting relationships between individuals, which has further deteriorating effects on population dynamics. In addition, several West African primate species form multi-species groups. These associations serve as predator detection and avoidance mechanisms, in which the Diana monkey seems to play a crucial role (Bshary and Noe 1997, Noe and Bshary 1997). Decimation of individual species that are especially vulnerable to hunting can therefore have further far-reaching fatal effects on a wide range of additional species. Of the two nocturnal primates, only the Potto is hunted, but only on a very small scale.

Habitat fragmentation is another major factor that leads to small, isolated populations. These populations being socially and genetically isolated are especially threatened in their survival by diseases, inbreeding and other factors.

Furthermore, habitat fragmentation severely limits the movement of primate groups and thereby their possibility of migration in case of further habitat destruction. In the case of chimpanzees, habitat destruction and fragmentation increases the frequency of inter-group fighting that can take a heavy toll on their populations.

In all four forest reserves surveyed during this RAP, primate populations are under serious threat from hunting, habitat destruction and fragmentation, and agricultural activities (Krokosua Hills). While most hunters use baikal shotguns and hunt all diurnal monkeys, only a small minority of hunters report hunting with dogs and therefore not hunting primates at all. Interestingly most of the poachers living in the villages close to the reserves seem to be immigrants either from the Greater Accra region, Central, Eastern or the three northern regions of Ghana. These people consistently report that they wouldn't kill chimpanzees since these animals are to close to humans. However, the same hunters report that the migrant population eats chimpanzees, that hunters of these tribes visit the forests regularly (sometimes from towns far away), that they hunt chimpanzees, and that there is a market for chimpanzee meat in all the bigger towns around the reserves. One hunter assured us that there is still a demand for young chimpanzees as pets and that even zoos in bigger towns would be willing to pay considerable amounts of money for young chimpanzees. Other hunting techniques like snares, although not primarily directed towards primates, pose a serious threat to ground-dwelling primates like chimpanzees and mangabeys. Although chimpanzees are known to be able to free themselves from wire traps, they often lose the affected limb or die from resulting bacterial infection (Goodall 1986, Boesch and Boesch-Achermann 2000).

These reports are in accordance with the low observation

rate of all diurnal primates in the four forest reserves. The low encounter rate might be, to some extent, explained by behavioral adaptations to the high hunting pressure resulting in a more cryptic lifestyle and a lower vocalization rate. The fact that species that are the most conspicuous, like the Red Colobus monkey, the Roloway Monkey, the White-naped Mangabey and the White-thighed Colobus, have not been found in this survey strongly indicates that intense hunting pressure leading to reduced population size if not local extinction is mainly responsible for our results. However, other reports conducted during the last years have confirmed the presence of threatened species in the surveyed forest reserves. At Draw River the presence of the Roloway Monkey and the White-naped Mangabey was confirmed (Abedi-Lartey 1998), at Boi-Tano the presence of the White-naped Mangabey (Abedi-Lartey 1999), and at Krokosua Hills the presence of the Roloway Monkey, the White-naped Mangabey and the White-thighed Colobus was confirmed during surveys conducted by various researchers (Magnusson in press, Abedi-Lartey 1999).

Although our results failed to confirm the presence of several primate species in the four forest reserves, we could confirm the presence of chimpanzees at Draw River and Tano Nimiri. At Draw River, chimpanzee presence has been claimed by local hunters (Abedi-Lartey 1998) but until now critical evidence was missing. Together with other reports that confirm the presence of chimpanzees at Tano Nimiri and Krokosua Hills (Abedi-Lartey 1998, 1999), this means that the visited forest patches still harbor chimpanzee populations although possibly in relatively low numbers. Given the fact that the total number of chimpanzees in Ghana is currently estimated to be less than 500 (Teleki 1989, Magnusson et al. 2003), the protection of every confirmed chimpanzee population should have highest priority. First, protection of these populations might be crucial to the survival of this species in Ghana. Second, the chimpanzee as a flag-ship species might allow the attraction of donors to fund protection measures in these forest reserves and thereby increase the probability of the survival of less charismatic threatened species within these areas. Finally, a healthy and well protected chimpanzee population opens up the possibility of habituating a group for ecotourism purposes, which could help improve revenues for the local population and thereby make wildlife protection much more attractive.

Although all four forest reserves were subject to severe human disturbance due to poaching, logging and illegal farming, they still hold the potential of being adequate primate habitats. In particular the Draw River GSBA area appears to consist of a relatively undisturbed forest canopy and its border with the Ankasa National Park makes it an especially interesting area for conservation. The Boi-Tano and Tano Nimiri reserves appear to contain a small chimpanzee population and hunters have recently reported sighting two Red Colobus monkeys. However the GSBA areas outlined for these two forest reserves are located in the northern part of Tano Nimiri and the southern part of Boi-Tano, while the areas in between are scheduled to be logged in the near future. Logging of these parts would not only add to further fragmentation of valuable primate habitat, but destroy the forest structure of an area where chimpanzee presence was reliably confirmed. Given the already precarious state of this species in Ghana, this has to be avoided by all means. Of all reserves surveyed, the lowlands of Krokosua Hills seemed to be the most disturbed area in terms of illegal farming and logging activities. The fact that a village is located within the GSBA area of the reserve adds to the problematic status of this forest block. However, various other surveys conducted in this area reported the presence of several threatened primate species and awareness campaigns to promote wildlife conservation have already begun. Furthermore, this reserve is already monitored by staff of the Wildlife Division of the Bia National Park.

In summary we conclude that all four forest reserves still have the potential to be adequate primate habitats. However urgent protection and awareness creation measures have to be implemented as soon as possible to secure these forest fragments as critical habitats for globally threatened species, such as the Western Chimpanzee, and prevent such crucial and highly endangered Ghanaian primate species from local extinction.

## CONSERVATION RECOMMENDATIONS

- Create biological corridors to connect the different reserves with neighboring national parks, and upgrade the protection status of the reserves accordingly. This would ensure that additional habitat critical for the threatened species is secured.

- Prevent logging, especially that which leads to further habitat fragmentation and destruction of confirmed chimpanzee habitat (Tano Nimiri and Boi-Tano).

- Establish wildlife guard camps in all areas and train guards in patrolling techniques. Regular monitoring of guard activities is necessary, as well as continuous motivation and support.

- Establish a conservation education and awareness program for the different areas. Chimpanzees, due to their acknowledged closeness to humans, could play a vital role in these educational campaigns.

- Strictly monitor illegal farming and logging activities within the reserve borders (especially at Krokosua Hills). Illegal farms should not only be removed, but farmers should be fined for forest destruction. Immigration into the Mmem village, which is within the forest reserve, is high and ongoing, leading to an increase in demand for land and consequently to a high conflict potential between villagers and wildlife guards. Therefore, in the long run, a relocation of the Mmem village has to be seriously considered.

- Further surveys are strongly recommended at sites where hunters have repeatedly claimed recent sightings of Miss Waldron's Red Colobus (Boi-Tano, Tano Nimiri and Krokosua Hills). Confirmation of the presence of this monkey species in these areas could drastically speed up and facilitate the elevation of protection status and the acquisition of financial support from conservation agencies.

- Attract international scientists and local students to establish long-term research sites in these areas. A growing body of experience clearly indicates that the permanent presence of researchers is the best means to secure protection of animals and reduce poaching activities.

## REFERENCES

Abedi-Lartey, M. 1998. Survey of endangered Forest Primates in Western Ghana. Paper presented at the XVIIth International Primatological Congress, Antananarivo, Madagascar.

Abedi-Lartey, M. 1999. Survey of endangered endemic Primates in Western Ghana. Report submitted to Wildlife Conservation Society, New York, and the Wildlife Department, Accra, Ghana.

Bearder, S.K. 1987. Lorises, bushbabies and tarsiers: Diverse societies in solitary foragers. *In*: Smuts, B.B., D.L. Cheney, R.M. Seyfarth, R.W. Wrangham and T.T. Struhsaker (eds.). Primate Societies. Pp. 11-24. Chicago: University of Chicago Press.

Boesch, C. 1991. The effects of leopard predation on grouping patterns in forest chimpanzees. Behaviour 117: 220-242.

Boesch, C. and H. Boesch. 1989. Hunting behaviour of wild chimpanzees in the Taï National Park. American Journal of Physical Anthropology 78:547-573.

Boesch, C. and H. Boesch-Achermann. 2000. The Chimpanzees of the Taï Forest: Behavioural Ecology and Evolution. Oxford University Press, Oxford.

Bshary, R. and R. Noe. 1997. Red colobus and Diana monkeys provide mutual protection against predators. Animal Behaviour. 54: 1461-1474.

Butynski, T.M. 2001. Africa's great apes. *In*: Beck, B., T.S. Stoinski, M.Hutchins, T.L. Maple, B.Norton, A.Rowan, E.F.Stevens and A. Arluke (eds.). Great apes and Humans: The Ethics of Coexistence, pp. 3-56. Washington, DC: Smithsonian Institution Press.

Caspary, H.-U., I. Koné, C. Prouot and M. de Pauw. 2001. La chasse et la filière viande debrousse dans l'espace Taï, Côte d'Ivoire.

Chapman, C.A. 1995. Primate seed dispersal: Coevolution and conservation implications. Evolutionary Anthropology 4:74-82.

Chapman, C.A. and D. Onderdonk. 1998. Forests without primates: Primate/plant codependency. American Journal of Primatology 45: 127-141.

Cheney, D.L. and R.M. Seyfarth. 1990. How monkeys see the world. Chicago: University of Chicago Press.

Cowlishaw, G. 1994. Vulnerability to predation in baboon populations. Behaviour 131: 293-304.

Cowlishaw, G. and R.I.M. Dunbar. 2000. Primate conservation biology. Chicago: University of Chicago Press.

Goodall, J. 1986. The Chimpanzees of Gombe. Belknap Press, Harvard University, Cambridge, MA.

Hausfater, G. 1976. Predatory behaviour of yellow baboons. Behaviour 56: 44-68.

Magnusson, L., M. Adu-Nsiah and D. Kpelle. 2003. Chapter 13: Ghana. Pp 111-116. *In*: Kormos, R., C. Boesch, M. Bakarr and T. Butynski (eds.). Status Survey and Conservation Action Plan: West African Chimpanzee. IUCN/SSC Action Plan. Gland, Switzerland, IUCN.

Magnusson, L. In press. Distribution and abundance of the Roloway monkey *Cercopithecus diana roloway*, and other Primate species in Ghana. African Primates.

Mittermeier, R.A. and D.L. Cheney. 1987. Conservation of Primates and their habitats. *In*: Smuts, B.B., D.L. Cheney, R.M. Seyfarth, R.W. Wrangham and T.T. Struhsaker (eds.). Primate Societies. Pp. 11-24. Chicago: University of Chicago Press.

Mittermeier, R.A., N. Myers and J.B. Thomsen. 1998. Biodiversity hotspots and major tropical wilderness areas: approaches to setting conservation priorities. Conservation Biology, 12:516-520.

Noe, R.and R. Bshary. 1997. The formation of red colobus-diana monkey associations under predation pressure from chimpanzees. Proceedings of the Royal Society of London - Series B: Biological Sciences. 264: 253-259.

Oates, J.F., T.T. Struhsaker and G.H. Whitesides. 1996. Extinction Faces Ghana's Red Colobus monkey and other locally endemic subspecies. Primate Conservation 17: 138-144.

Oates, J.F., M. Abedi-Lartey, W.S. McGraw, T.T. Struhsaker and G.H. Whitesides. 2000. Extinction of a West African Red Colobus Monkey. Conservation Biology 14: 1526-1532.

Struhsaker, T.T. and M. Leakey. 1990. Prey selectivity by crowned hawk-eagles onmonkeys in the Kibale Forest, Uganda. Behavioral Ecology and Sociobiology 26:435-443.

Teleki, G. 1989. Population status of wild chimpanzees (*Pan troglodytes*) and threats to survival. *In*: Heltne, P. and L. Marquardt (eds.). Understanding chimpanzees. Pp. 312-353. Havard University Press, Cambridge, MA.

Terborgh, J. 1983. Five New World primates: A study in comparative ecology. Princeton: Princeton University Press.

# Appendix 1

## Summary of the composition of plant biodiversity in the study area, categorized by family, life-form or habit, ecological guild and conservation rating

*Patrick Ekpe*

Species in bold are rated as Black Star species.

| Family | Scientific Name | Habit | Guild | Star | Draw River | Boi-Tano | Krokosua |
|--------|-----------------|-------|-------|------|------------|----------|----------|
| Acanthaceae | *Asystasia vogeliana* | Herb | Pioneer | Green | 1 | 0 | 0 |
| Acanthaceae | *Brillantaisia owariensis* | Herb | Pioneer | Green | 1 | 0 | 1 |
| Acanthaceae | *Elytraria marginata* | Herb | Shade | Green | 0 | 0 | 1 |
| Acanthaceae | *Lankesteria brevior* | Herb | | Green | 1 | 1 | 0 |
| Adiantaceae | *Adiantum vogelii* | Herb/Fern | | Green | 1 | 0 | 0 |
| Adiantaceae | *Pteris burtonii* | Herb/Fern | | Green | 1 | 0 | 0 |
| Agavaceae | *Dracaena adamii* | Shrub | Shade | Gold | 1 | 1 | 0 |
| Agavaceae | *Dracaena camerooniana* | Shrub | Shade | Green | 1 | 1 | 1 |
| Agavaceae | *Dracaena cerasifera* | Shrub | Shade | Blue | 1 | 1 | 0 |
| Agavaceae | *Dracaena mannii* | Shrub | | Green | 1 | 1 | 1 |
| Agavaceae | *Dracaena phrynoides* | Shrub | Shade | Green | 1 | 1 | 1 |
| Agavaceae | *Dracaena surculosa* | Shrub | Shade | Green | 1 | 1 | 1 |
| Amaryllidaceae | *Scadoxus cinnabarinus* | Herb | | Green | 1 | 0 | 1 |
| Anacardiaceae | *Antrocaryon micraster* | Tree | NPLD | Red | 0 | 0 | 1 |
| Anacardiaceae | *Lannea welwitschii* | Tree | Pioneer | Green | 1 | 1 | 1 |
| Anacardiaceae | *Pseudospondias microcarpa* | Tree | Swamp | Green | 0 | 0 | 1 |
| Anacardiaceae | *Spondis mombin* | Tree | | Green | 0 | 0 | 1 |
| Anacardiaceae | *Trichoscypha arborea* | Tree | Shade | Green | 1 | 1 | 0 |
| Anacardiaceae | *Trichoscypha bijuga* | Tree | Shade | Gold | 1 | 0 | 0 |
| Anacardiaceae | *Trichoscypha lucens* | Tree | | Gold | 1 | 1 | 1 |
| Anacardiaceae | *Trichoscypha mannii* | Tree | | Gold | 1 | 1 | 0 |
| Ancistrocladaceae | *Ancistrocladus guineensis* | Liane/Climber | | Blue | 0 | 1 | 0 |
| Annonaceae | *Annickia polycarpa* | Tree | Shade | Green | 1 | 1 | 1 |
| Annonaceae | *Anonidium mannii* | Tree | Shade | Blue | 1 | 0 | 1 |
| Annonaceae | *Cleistopholis patens* | Tree | Pioneer | Green | 1 | 1 | 1 |
| Annonaceae | *Dennettia tripetala* | Tree | Shade | Gold | 0 | 0 | 1 |
| Annonaceae | *Friesodielsia enghiana* | Liane/Climber | | Green | 1 | 1 | 0 |
| Annonaceae | *Friesodielsia velutina* | Liane/Climber | | Green | 1 | 1 | 1 |
| Annonaceae | *Greenyodendron oliveri* | Tree | Shade | Green | 1 | 1 | 1 |
| Annonaceae | *Hexalobus crispiflorus* | Tree | Shade | Green | 1 | 0 | 1 |

| Family | Scientific Name | Habit | Guild | Star | Draw River | Boi-Tano | Krokosua |
|---|---|---|---|---|---|---|---|
| **Annonaceae** | *Isolona deightonii* | **Tree** | **Shade** | **Black** | **1** | **1** | **0** |
| Annonaceae | *Mischogyne elliotiana* | Tree | Shade | Blue | 0 | 0 | 1 |
| Annonaceae | *Monanthotaxis laurentii* | Shrub | | Green | 0 | 0 | 1 |
| Annonaceae | *Monodora myristica* | Tree | Shade | Green | 1 | 1 | 1 |
| Annonaceae | *Monodora tenuifolia* | Tree | Pioneer | Green | 1 | 1 | 1 |
| Annonaceae | *Piptostigma fasciculatum* | Tree | Shade | Green | 1 | 0 | 1 |
| Annonaceae | *Piptostigma fugax* | Tree | Shade | Gold | 1 | 1 | 0 |
| Annonaceae | *Polyceratocarpus parviflorus* | Tree | Shade | Blue | 0 | 0 | 1 |
| Annonaceae | *Uvaria afzelii* | Liane/Climber | NPLD | Green | 1 | 0 | 0 |
| Annonaceae | *Uvaria mocoli* | Liane/Climber | NPLD | Green | 1 | 0 | 1 |
| Annonaceae | *Uvariastrum pierreanum* | Tree | Shade | Green | 0 | 0 | 1 |
| Annonaceae | *Uvariodendron calophyllum* | Tree | Shade | Green | 0 | 0 | 1 |
| Annonaceae | *Uvariopsis globiflora* | Tree | Shade | Blue | 0 | 0 | 1 |
| Annonaceae | *Xylopia acutiflora* | Liane/Climber | NPLD | Green | 1 | 1 | 0 |
| Annonaceae | *Xylopia elliotii* | Tree | Shade | Gold | 1 | 1 | 1 |
| Annonaceae | *Xylopia quintasii* | Tree | Shade | Green | 1 | 1 | 0 |
| Annonaceae | *Xylopia rubescens* | Tree | Swamp | Gold | 1 | 1 | 0 |
| Annonaceae | *Xylopia sp. aff. hypolampra* | Tree | Shade | Gold | 1 | 1 | 0 |
| Annonaceae | *Xylopia staudtii* | Tree | Shade | Green | 1 | 1 | 0 |
| Annonaceae | *Xylopia villosa* | Tree | Shade | Green | 1 | 1 | 1 |
| Apocynaceae | *Alafia barteri* | Liane/Climber | NPLD | Green | 1 | 1 | 1 |
| Apocynaceae | *Alafia lucida* | Liane/Climber | NPLD | Blue | 1 | 1 | 0 |
| Apocynaceae | *Alafia schumannii* | Liane/Climber | | Green | 1 | 1 | 0 |
| Apocynaceae | *Alafia whytei* | Liane/Climber | | Gold | 1 | 0 | 0 |
| Apocynaceae | *Alstonia boonei* | Tree | Pioneer | Green | 0 | 1 | 1 |
| Apocynaceae | *Ancylobotrys scandens* | Liane/Climber | | Green | 1 | 0 | 0 |
| Apocynaceae | *Baissea baillonii* | Liane/Climber | | Green | 1 | 1 | 1 |
| Apocynaceae | *Baissea leonensis* | Liane/Climber | NPLD | Blue | 1 | 1 | 0 |
| Apocynaceae | *Callichilia subsessilis* | Liane/Climber | | Blue | 1 | 0 | 0 |
| Apocynaceae | *Farquharia elliptica* | Liane/Climber | NPLD | Green | 0 | 1 | 0 |
| Apocynaceae | *Funtumia africana* | Tree | | Green | 1 | 1 | 1 |
| Apocynaceae | *Funtumia elastica* | Tree | | Pink | 0 | 1 | 1 |
| Apocynaceae | *Hunteria umbellata* | Tree | Shade | Green | 1 | 1 | 1 |
| Apocynaceae | *Landolphia calabarica* | Liane/Climber | NPLD | Green | 1 | 1 | 1 |
| Apocynaceae | *Landolphia dulcis* | Liane/Climber | NPLD | Green | 1 | 1 | 0 |
| Apocynaceae | *Landolphia forentiana* | Liane/Climber | | Blue | 1 | 1 | 1 |
| Apocynaceae | *Landolphia hirsuta* | Liane/Climber | | Blue | 1 | 1 | 1 |
| Apocynaceae | *Landolphia incerta* | Liane/Climber | | Green | 1 | 1 | 1 |
| **Apocynaceae** | *Landolphia membranacea* | **Liane/Climber** | | **Black** | **0** | **1** | **0** |
| Apocynaceae | *Landolphia micrantha* | Liane/Climber | NPLD | Blue | 1 | 1 | 0 |
| Apocynaceae | *Landolphia owariensis* | Liane/Climber | | Green | 1 | 1 | 1 |

| Family | Scientific Name | Habit | Guild | Star | Draw River | Boi-Tano | Krokosua |
|--------|-----------------|-------|-------|------|------------|----------|----------|
| Apocynaceae | *Motandra guineensis* | Liane/Climber | NPLD | Green | 0 | 0 | 1 |
| Apocynaceae | *Pleiocarpa mutica* | Tree | Shade | Green | 1 | 1 | 1 |
| Apocynaceae | *Pleioceras barteri* | Shrub | Pioneer | Green | 1 | 0 | 0 |
| Apocynaceae | *Rauvolfia mannii* | Shrub | Shade | Blue | 1 | 0 | 1 |
| Apocynaceae | *Strophanthus gratus* | Liane/Climber | | Pink | 1 | 0 | 0 |
| Apocynaceae | *Strophanthus hispidus* | Liane/Climber | Pioneer | Pink | 0 | 1 | 1 |
| Apocynaceae | *Tabernaemontana africana* | Tree | Shade | Green | 0 | 0 | 1 |
| Apocynaceae | *Voacanga africana* | Tree | Pioneer | Green | 1 | 0 | 1 |
| Apocynaceae | *Voacanga bracteata* | Shrub | Shade | Green | 1 | 0 | 0 |
| Araceae | *Anchomanes difformis* | Herb | Pioneer | Green | 0 | 1 | 1 |
| Araceae | *Cercestis afzelii* | Liane/Climber | Shade | Green | 1 | 1 | 1 |
| Araceae | *Cercestis dinklagei* | Liane/Climber | | Green | 1 | 1 | 1 |
| Araceae | *Cercestis ivorensis* | Liane/Climber | Shade | Blue | 1 | 1 | 0 |
| Araceae | *Culcasia angolensis* | Herb | NPLD | Green | 1 | 1 | 1 |
| Araceae | *Culcasia liberica* | Herb | | Gold | 1 | 1 | 0 |
| Araceae | *Culcasia parviflora* | Herb | | Green | 1 | 0 | 0 |
| Araceae | *Culcasia scandens* | Herb | Shade | Green | 1 | 1 | 1 |
| Araceae | *Culcasia striolata* | Herb | Shade | Green | 1 | 1 | 1 |
| Araceae | *Nephthytis afzelii* | Herb | Shade | Blue | 0 | 0 | 1 |
| **Araceae** | ***Nephthytis swainei*** | **Herb** | **Shade** | **Black** | **0** | **1** | **0** |
| Araceae | *Rhaphiodophora africana* | Herb | Shade | Green | 0 | 0 | 1 |
| Araliaceae | *Schefflera barteri* | Liane/Climber | | Blue | 1 | 0 | 0 |
| Aristolochiaceae | *Pararistolochia goldiena* | Liane/Climber | | Blue | 0 | 0 | 1 |
| Asclepiadaceae | *Gongronema latifolium* | Liane/Climber | | Green | 0 | 0 | 1 |
| Asclepiadaceae | *Pergularia daemia* | Liane/Climber | Pioneer | Green | 0 | 0 | 1 |
| Asclepiadaceae | *Tylophora conspicua* | Liane/Climber | Pioneer | Green | 0 | 0 | 1 |
| Aspidiaceae | *Ctenitis jenseniae* | Herb/Fern | | Green | 1 | 0 | 0 |
| Aspidiaceae | *Ctenitis lanigera* | Herb/Fern | | Green | 1 | 0 | 0 |
| Aspidiaceae | *Ctenitis protensa* | Herb/Fern | | Green | 1 | 0 | 0 |
| Aspleniaceae | *Asplenium africanum* | Herb/Fern | | Green | 1 | 0 | 1 |
| Asteraceae | *Ageratum conyzoides* | Herb | Pioneer | Green | 0 | 0 | 0 |
| Asteraceae | *Chromolaena odorata* | Herb | Pioneer | Green | 0 | 0 | 1 |
| Balanophoraceae | *Thonningia sanguinea* | Herb | Shade | Green | 1 | 0 | 1 |
| Begoniaceae | *Begonia macrocarpa* | Herb | Shade | Blue | 0 | 0 | 1 |
| Bignonaceae | *Kigelia africana* | Tree | NPLD | Green | 1 | 0 | 1 |
| Bignonaceae | *Newbouldia laevis* | Tree | Pioneer | Green | 1 | 0 | 1 |
| Bignonaceae | *Sterospermum acuminatissimum* | Tree | Pioneer | Green | 0 | 0 | 1 |
| Bombacaceae | *Ceiba pentandra* | Tree | Pioneer | Pink | 1 | 1 | 1 |
| Bombacaceae | *Rhodognaphalon brevicuspe* | Tree | Pioneer | Red | 1 | 1 | 1 |
| Burseraceae | *Canarium schweinfurthii* | Tree | Pioneer | Pink | 1 | 1 | 1 |

A biological assessment of the terrestrial ecosystems of the Draw River, Boi-Tano, Tano Nimiri and Krokosua Hills forest reserves, southwestern Ghana

81

| Family | Scientific Name | Habit | Guild | Star | Draw River | Boi-Tano | Krokosua |
|--------|-----------------|-------|-------|------|------------|----------|----------|
| Burseraceae | *Dacryodes klaineana* | Tree | Shade | Green | 1 | 1 | 1 |
| Caesalpiniaceae | *Afzelia bella* | Tree | NPLD | Red | 0 | 0 | 0 |
| Caesalpiniaceae | *Amphimas pterocarpoides* | Tree | NPLD | Green | 1 | 1 | 1 |
| Caesalpiniaceae | *Anthonotha fragrans* | Tree | NPLD | Green | 1 | 1 | 1 |
| Caesalpiniaceae | *Anthonotha macrophylla* | Tree | Shade | Green | 1 | 1 | 1 |
| **Caesalpiniaceae** | ***Anthonotha sassandraensis*** | **Tree** | **Shade** | **Black** | **1** | **1** | **0** |
| Caesalpiniaceae | *Berlinia confusa* | Tree | Shade | Green | 1 | 1 | 0 |
| **Caesalpiniaceae** | ***Berlinia occidentalis*** | **Tree** | **Swamp** | **Black** | **1** | **1** | **0** |
| Caesalpiniaceae | *Berlinia tomentella* | Tree | Shade | Blue | 1 | 1 | 0 |
| Caesalpiniaceae | *Bussea occidentalis* | Tree | NPLD | Green | 1 | 1 | 1 |
| Caesalpiniaceae | *Childlowia sanguinea* | Tree | Shade | Blue | 0 | 0 | 1 |
| Caesalpiniaceae | *Copaifera salikounda* | Tree | Shade | Red | 1 | 1 | 1 |
| Caesalpiniaceae | *Crudia ganonensis* | Tree | Shade | Gold | 1 | 0 | 0 |
| Caesalpiniaceae | *Cynometra ananta* | Tree | Shade | Pink | 1 | 1 | 0 |
| Caesalpiniaceae | *Daniellia ogea/thurifera* | Tree | Pioneer | Red | 1 | 1 | 0 |
| Caesalpiniaceae | *Daniellia thurifera* | Tree | Pioneer | Pink | 1 | 1 | 1 |
| Caesalpiniaceae | *Dialium aubrevillei* | Tree | Shade | Green | 1 | 1 | 1 |
| Caesalpiniaceae | *Dialium dinklaigei* | Tree | NPLD | Green | 1 | 1 | 1 |
| Caesalpiniaceae | *Dialium guineense* | Tree | non-forest | Green | 0 | 0 | 1 |
| Caesalpiniaceae | *Didelotia idae* | Tree | Swamp | Gold | 1 | 0 | 0 |
| Caesalpiniaceae | *Didelotia unifoliolata* | Tree | Shade | Gold | 1 | 0 | 0 |
| Caesalpiniaceae | *Distemonanthus benthamianus* | Tree | NPLD | Pink | 1 | 1 | 1 |
| Caesalpiniaceae | *Duparquetia orchidacea* | Liane/Climber | NPLD | Green | 1 | 1 | 0 |
| Caesalpiniaceae | *Erythrophleum ivorense* | Tree | NPLD | Pink | 1 | 1 | 0 |
| Caesalpiniaceae | *Gilbertiodendron preussii* | Tree | | Green | 1 | 0 | 0 |
| **Caesalpiniaceae** | ***Gilbertiodendron splendidum*** | **Tree** | | **Black** | **0** | **1** | **0** |
| Caesalpiniaceae | *Griffonia simplicifolia* | Liane/Climber | NPLD | Green | 1 | 1 | 1 |
| Caesalpiniaceae | *Guibourtia ehia* | Tree | NPLD | Red | 0 | 0 | 1 |
| Caesalpiniaceae | *Hymenostegia afzelii* | Tree | Shade | Green | 1 | 0 | 1 |
| Caesalpiniaceae | *Hymenostegia aubrevillei* | Tree | Shade | Gold | 0 | 0 | 1 |
| **Caesalpiniaceae** | ***Hymenostegia gracilipes*** | **Tree** | **Shade** | **Black** | **1** | **1** | **0** |
| Caesalpiniaceae | *Stemonocoleus micranthus* | Shrub | NPLD | Blue | 0 | 0 | 1 |
| Capparaceae | *Buchholzia coriacea* | Tree | Shade | Green | 1 | 1 | 1 |
| Capparaceae | *Euadenia eminens* | Shrub | | Blue | 1 | 0 | 1 |
| Capparaceae | *Euadenia trifoliolata* | Shrub | Shade | Blue | 1 | 1 | 1 |
| Celastraceae | *Apodostigma pallens* | Liane/Climber | NPLD | Green | 1 | 0 | 0 |
| Celastraceae | *Campylostemon angolense* | Liane/Climber | | Green | 0 | 0 | 1 |
| Celastraceae | *Cuervea macrophylla* | Liane/Climber | NPLD | Green | 1 | 0 | 1 |
| Celastraceae | *Salacia columna* | Liane/Climber | Shade | Green | 1 | 1 | 0 |
| Celastraceae | *Salacia cornifolia* | Liane/Climber | | Green | 1 | 0 | 0 |

| Family | Scientific Name | Habit | Guild | Star | Draw River | Boi-Tano | Krokosua |
|---|---|---|---|---|---|---|---|
| Celastraceae | *Salacia elegans* | Liane/Climber | NPLD | Green | 0 | 1 | 1 |
| Celastraceae | *Salacia erecta* | Liane/Climber | NPLD | Green | 1 | 0 | 0 |
| Celastraceae | *Salacia miegei* | Liane/Climber | NPLD | Blue | 1 | 1 | 0 |
| Celastraceae | *Salacia nitida* | Liane/Climber | | Green | 1 | 0 | 0 |
| Celastraceae | *Salacia pallescens* | Liane/Climber | | Green | 0 | 1 | 1 |
| Celastraceae | *Salacia preussii* | Liane/Climber | | Green | 0 | 0 | 1 |
| Celastraceae | *Salacia pyriformis* | Liane/Climber | NPLD | Green | 1 | 0 | 0 |
| Celastraceae | *Salacia zenkeri* | Liane/Climber | | Blue | 1 | 0 | 0 |
| Celastraceae | *Simicratea welwitschii* | Liane/Climber | | Green | 0 | 0 | 1 |
| Chrysobalanaceae | *Dactyladenia barteri* | Tree | Shade | Blue | 1 | 1 | 0 |
| Chrysobalanaceae | *Maranthes aubrevillei* | Tree | Shade | Gold | 1 | 1 | 0 |
| Chrysobalanaceae | *Maranthes chrysophylla* | Tree | Shade | Blue | 1 | 1 | 0 |
| Chrysobalanaceae | *Maranthes glabra* | Tree | Shade | Green | 1 | 1 | 0 |
| Chrysobalanaceae | *Parinari excelsa* | Tree | NPLD | Green | 1 | 1 | 1 |
| Combretaceae | *Combretum mucronatum* | Liane/Climber | Pioneer | Green | 0 | 0 | 1 |
| Combretaceae | *Strephonema pseudocola* | Tree | Shade | Blue | 1 | 1 | 0 |
| Combretaceae | *Terminalia ivorensis* | Tree | Pioneer | Scarlet | 0 | 0 | 1 |
| Combretaceae | *Terminalia superba* | Tree | Pioneer | Pink | 1 | 0 | 1 |
| Commelinaceae | *Aneilema aequinoctiale* | Herb | Pioneer | Green | 0 | 1 | 0 |
| Commelinaceae | *Bufforrestia obovata* | Herb | | Gold | 1 | 1 | 0 |
| Commelinaceae | *Coeleotrype laurentii* | Herb | | Green | 1 | 0 | 0 |
| Commelinaceae | *Commelina capitata* | Herb | Pioneer | Green | 1 | 0 | 0 |
| Commelinaceae | *Commelina macrosperma* | Herb | Pioneer | Green | 0 | 1 | 1 |
| Commelinaceae | *Floscopa africana* | Herb | | Green | 1 | 0 | 0 |
| Commelinaceae | *Palisota barteri* | Herb | Pioneer | Green | 1 | 0 | 1 |
| Commelinaceae | *Palisota hirsuta* | Herb | Pioneer | Green | 1 | 0 | 1 |
| Commelinaceae | *Pollia condensata* | Herb | Pioneer | Green | 1 | 1 | 0 |
| Commelinaceae | *Polyspatha paniculata* | Herb | Pioneer | Green | 1 | 0 | 0 |
| Commelinaceae | *Stanfieldiella imperforata* | Herb | | Green | 0 | 0 | 1 |
| Connaraceae | *Agelaea paradoxa* | Liane/Climber | NPLD | Green | 1 | 1 | 1 |
| Connaraceae | *Agelaea paradoxa var. paradoxa* | Liane/Climber | NPLD | Green | 1 | 0 | 1 |
| Connaraceae | *Agelaea pentagyna* | Liane/Climber | Shade | Green | 1 | 1 | 1 |
| Connaraceae | *Cnestis bomiensis* | Liane/Climber | | Gold | 0 | 1 | 0 |
| Connaraceae | *Cnestis corniculata* | Liane/Climber | NPLD | Green | 1 | 1 | 1 |
| Connaraceae | *Cnestis ferruginea* | Liane/Climber | Pioneer | Green | 1 | 1 | 0 |
| Connaraceae | *Connarus africanus* | Shrub | NPLD | Green | 1 | 1 | 0 |
| **Connaraceae** | ***Connarus thonningii*** | **Liane/Climber** | | **Black** | **1** | **0** | **0** |
| Connaraceae | *Manotes expansa* | Liane/Climber | | Green | 1 | 1 | 0 |
| Connaraceae | *Rourea coccinea* | Liane/Climber | Pioneer | Green | 0 | 0 | 1 |
| Convolvulaceae | *Calycobolus africanus* | Liane/Climber | Shade | Green | 1 | 1 | 1 |

A biological assessment of the terrestrial ecosystems of the Draw River, Boi-Tano,
Tano Nimiri and Krokosua Hills forest reserves, southwestern Ghana

83

| Family | Scientific Name | Habit | Guild | Star | Draw River | Boi-Tano | Krokosua |
|---|---|---|---|---|---|---|---|
| Cucurbitaceae | *Coccinea barteri* | Liane/Climber | Pioneer | Green | 0 | 0 | 1 |
| **Cucurbitaceae** | ***Momordica sylvatica*** | **Liane/Climber** | **Pioneer** | **Black** | **0** | **1** | **0** |
| Cyatheaceae | *Cyathea manniana* | Herb/Fern | Swamp | Blue | 0 | 1 | 0 |
| Cyperaceae | *Mapania baldwinii* | Herb | | Blue | 0 | 1 | 1 |
| Davalliaceae | *Arthropteris palisotii* | Herb/Fern | | Green | 1 | 0 | 0 |
| Davalliaceae | *Nephrolepis bisserata* | Herb/Fern | Shade | Green | 1 | 0 | 0 |
| Dichapetalaceae | *Dichapetalum angolense* | Liane/Climber | NPLD | Green | 1 | 1 | 0 |
| Dichapetalaceae | *Dichapetalum crassifolium* | Liane/Climber | NPLD | Green | 0 | 1 | 1 |
| **Dichapetalaceae** | ***Dichapetalum filicaule*** | **Liane/Climber** | **Shade** | **Black** | **1** | **1** | **0** |
| Dichapetalaceae | *Dichapetalum heudelotii* var. *heudelotii* | Liane/Climber | Shade | Green | 1 | 1 | 0 |
| Dichapetalaceae | *Dichapetalum heudelotii* var. *ndongense* | Liane/Climber | Shade | Green | 1 | 0 | 0 |
| Dichapetalaceae | *Dichapetalum johnstonii* | Liane/Climber | NPLD | Green | 1 | 1 | 0 |
| Dichapetalaceae | *Dichapetalum madagascariense* | Tree | Shade | Green | 0 | 0 | 1 |
| Dichapetalaceae | *Dichapetalum oblongum* | Shrub | Shade | Blue | 1 | 0 | 1 |
| Dichapetalaceae | *Dichapetalum pallidum* | Liane/Climber | Shade | Green | 1 | 1 | 1 |
| Dichapetalaceae | *Dichapetalum toxicarium* | Liane/Climber | NPLD | Gold | 1 | 1 | 0 |
| Dichapetalaceae | *Tapura ivorensis* | Tree | Shade | Black | 1 | 1 | 0 |
| Dilleniaceae | *Tetracera alnifolia* | Liane/Climber | | Green | 1 | 1 | 0 |
| Dioscoreaceae | *Dioscorea minutiflora* | Liane/Climber | Pioneer | Green | 0 | 0 | 1 |
| Dioscoreaceae | *Dioscorea smilacifolia* | Liane/Climber | Pioneer | Green | 0 | 1 | 1 |
| Ebenaceae | *Diospyros chevalieri* | Shrub | Shade | Gold | 1 | 1 | 0 |
| Ebenaceae | *Diospyros gabunensis* | Tree | Shade | Blue | 0 | 1 | 0 |
| Ebenaceae | *Diospyros kamerunensis* | Tree | Shade | Green | 1 | 1 | 1 |
| Ebenaceae | *Diospyros monbuttensis* | Tree | Shade | Green | 0 | 0 | 1 |
| Ebenaceae | *Diospyros sanza-minika* | Tree | Shade | Blue | 1 | 1 | 0 |
| Ebenaceae | *Diospyros vignei* | Tree | Shade | Blue | 1 | 1 | 0 |
| Ebenaceae | *Diospyros viridicans* | Tree | Shade | Green | 0 | 1 | 1 |
| Erythroxylaceae | *Erythroxylum mannii* | Tree | Pioneer | Green | 1 | 1 | 0 |
| Euphorbiaceae | *Alchornea cordifolia* | Tree | Pioneer | Green | 0 | 0 | 1 |
| Euphorbiaceae | *Alchornea floribunda* | Tree | Shade | Green | 0 | 1 | 1 |
| Euphorbiaceae | *Amanoa bracteosa* | Tree | Swamp | Gold | 0 | 1 | 0 |
| Euphorbiaceae | *Anthostema aubryanum* | Tree | Swamp | Blue | 0 | 1 | 0 |
| Euphorbiaceae | *Antidesma laciniatum* | Tree | Shade | Green | 1 | 0 | 1 |
| **Euphorbiaceae** | ***Antidesma oblonga*** | **Tree** | **Shade** | **Black** | **1** | **1** | **0** |
| Euphorbiaceae | *Argomuellera macrophylla* | Shrub | | Green | 0 | 0 | 1 |
| Euphorbiaceae | *Bridelia atroviridis* | Tree | Pioneer | Green | 1 | 0 | 0 |
| Euphorbiaceae | *Bridelia grandis* | Tree | Pioneer | Green | 1 | 1 | 1 |
| Euphorbiaceae | *Bridelia micrantha* | Tree | Pioneer | Green | 1 | 0 | 0 |
| Euphorbiaceae | *Crotonogyne manniana* | Shrub | Shade | Gold | 0 | 1 | 0 |

| Family | Scientific Name | Habit | Guild | Star | Draw River | Boi-Tano | Krokosua |
|--------|-----------------|-------|-------|------|------------|----------|----------|
| Euphorbiaceae | *Discoglypremna caloneura* | Tree | Pioneer | Green | 1 | 0 | 1 |
| Euphorbiaceae | *Drypetes afzelii* | Tree | Shade | Gold | 1 | 1 | 0 |
| Euphorbiaceae | *Drypetes aubrevillei* | Tree | Shade | Blue | 1 | 1 | 1 |
| Euphorbiaceae | *Drypetes aylmeri* | Tree | Shade | Blue | 1 | 1 | 0 |
| Euphorbiaceae | *Drypetes chevalieri* | Tree | Shade | Green | 1 | 0 | 1 |
| Euphorbiaceae | *Drypetes gilgiana* | Tree | Shade | Green | 1 | 1 | 1 |
| Euphorbiaceae | *Drypetes ivorensis* | Tree | Shade | Blue | 1 | 1 | 0 |
| Euphorbiaceae | *Drypetes leonensis* | Tree | Shade | Blue | 1 | 0 | 1 |
| Euphorbiaceae | *Drypetes parvifolia* | Tree | Shade | Green | 0 | 0 | 1 |
| Euphorbiaceae | *Drypetes pellegrinii* | Tree | Shade | Gold | 1 | 0 | 1 |
| Euphorbiaceae | *Drypetes principum* | Tree | Shade | Green | 1 | 1 | 1 |
| Euphorbiaceae | *Elaeophorbia grandifolia* | Tree | Pioneer | Green | 0 | 0 | 1 |
| Euphorbiaceae | *Erythrococca anomala* | Shrub | | Green | 0 | 0 | 1 |
| Euphorbiaceae | *Macaranga barteri* | Tree | Pioneer | Green | 1 | 1 | 0 |
| Euphorbiaceae | *Macaranga heterophylla* | Tree | Pioneer | Green | 1 | 1 | 0 |
| Euphorbiaceae | *Macaranga heurifolia* | Tree | Pioneer | Green | 1 | 1 | 0 |
| Euphorbiaceae | *Maesobtrya barteri* | Tree | Shade | Green | 1 | 1 | 0 |
| Euphorbiaceae | *Mallotus oppositifolius* | Tree | Shade | Green | 0 | 0 | 1 |
| Euphorbiaceae | *Manniophyton fulvum* | Liane/Climber | NPLD | Green | 1 | 1 | 1 |
| Euphorbiaceae | *Mareya micrantha* | Tree | Shade | Green | 1 | 1 | 1 |
| Euphorbiaceae | *Margaritaria discoidea* | Tree | Pioneer | Green | 1 | 0 | 1 |
| Euphorbiaceae | *Microdesmis puberula* | Tree | Shade | Green | 1 | 1 | 1 |
| Euphorbiaceae | *Mildbraedia paniculata* ssp. *occidentalis* | Tree | | Gold | 0 | 1 | 0 |
| Euphorbiaceae | *Mildbraediodendron excelsum* | Tree | | Gold | 0 | 1 | 0 |
| Euphorbiaceae | *Phyllanthus profusus* | Tree | Shade | Gold | 1 | 0 | 0 |
| Euphorbiaceae | *Protomegabaria stapfiana* | Tree | Shade | Blue | 1 | 1 | 0 |
| Euphorbiaceae | *Pycnocoma macrophyla* | Shrub | Shade | Green | 1 | 1 | 1 |
| Euphorbiaceae | *Ricinodendron heudelotii* | Tree | Shade | Green | 1 | 1 | 1 |
| Euphorbiaceae | *Spondiathus preussii* | Tree | Swamp | Green | 0 | 1 | 0 |
| Euphorbiaceae | *Tetrapleura tetraptera* | Tree | Pioneer | Green | 0 | 0 | 1 |
| Euphorbiaceae | *Tetrochidium didymostemon* | Tree | Pioneer | Green | 1 | 0 | 1 |
| Euphorbiaceae | *Uapaca cobisieri* | Tree | NPLD | Green | 1 | 1 | 0 |
| Euphorbiaceae | *Uapaca guineensis* | Tree | NPLD | Green | 1 | 1 | 0 |
| Euphorbiaceae | *Uapaca paludosa* | Tree | Swamp | Gold | 1 | 1 | 0 |
| Flacourtiaceae | *Caloncoba echinata* | Tree | Shade | Green | 1 | 1 | 0 |
| Flacourtiaceae | *Caloncoba gilgiana* | Tree | Pioneer | Green | 0 | 1 | 1 |
| **Flacourtiaceae** | ***Dasylepis assinensis*** | **Tree** | **Shade** | **Black** | **1** | **1** | **1** |
| Flacourtiaceae | *Dasylepis brevipedicellata* | Tree | Shade | Green | 0 | 0 | 1 |
| Flacourtiaceae | *Homalium africanum* | Tree | Swamp | Green | 1 | 0 | 0 |
| Flacourtiaceae | *Homalium letestui* | Tree | NPLD | Green | 1 | 1 | 1 |

A biological assessment of the terrestrial ecosystems of the Draw River, Boi-Tano, Tano Nimiri and Krokosua Hills forest reserves, southwestern Ghana

85

| Family | Scientific Name | Habit | Guild | Star | Draw River | Boi-Tano | Krokosua |
|--------|-----------------|-------|-------|------|------------|----------|----------|
| Flacourtiaceae | *Homalium longistylum* | Tree | NPLD | Gold | 1 | 1 | 1 |
| Flacourtiaceae | *Ophiobotrys zenkeri* | Tree | NPLD | Blue | 0 | 0 | 1 |
| Flacourtiaceae | *Scottellia klaineana* | Tree | Shade | Pink | 1 | 1 | 1 |
| Flagellariaceae | *Flagellaria guneensis* | Liane/Climber | | Green | 1 | 1 | 0 |
| Graminae | *Guaduella macrostachys* | Herb | Shade | Green | 1 | 0 | 0 |
| Graminae | *Leptaspis zeylanica* | Herb | Shade | Green | 1 | 1 | 1 |
| Graminae | *Olyra latifolia* | Grass | Shade | Green | 1 | 1 | 1 |
| Guttiferae | *Allanblackia parviflora* | Tree | Shade | Green | 1 | 1 | 1 |
| Guttiferae | *Garcinia afzelii* | Tree | Shade | Red | 0 | 0 | 1 |
| Guttiferae | *Garcinia epunctata* | Tree | Shade | Red | 1 | 1 | 0 |
| Guttiferae | *Garcinia gnetoides* | Tree | Shade | Blue | 1 | 1 | 0 |
| Guttiferae | *Garcinia kola* | Tree | Shade | Scarlet | 0 | 1 | 1 |
| Guttiferae | *Garcinia smeathmannii* | Tree | Shade | Green | 1 | 1 | 0 |
| Guttiferae | *Mammea africana* | Tree | Shade | Pink | 1 | 1 | 1 |
| Guttiferae | *Pentadesma butyracea* | Tree | Shade | Blue | 1 | 1 | 0 |
| Guttiferae | *Symphonia globulifera* | Tree | Swamp | Green | 1 | 1 | 1 |
| Hernandiaceae | *Illigera vespertilio* | Liane/Climber | | Gold | 1 | 0 | 0 |
| Humiriaceae | *Sacoglottis gabonensis* | Tree | Swamp | Blue | 1 | 0 | 0 |
| Hymenophyllaceae | *Trichomanes guineense* | Herb/Fern | | Green | 1 | 0 | 0 |
| **Icacinaceae** | ***Alsodeiopsis chippii*** | **Shrub** | **Shade** | **Black** | **1** | **1** | **0** |
| Icacinaceae | *Alsodeiopsis staudtii* | Shrub | Shade | Green | 1 | 1 | 0 |
| Icacinaceae | *Chlamydocarya macrocarpa* | Liane/Climber | Shade | Gold | 0 | 1 | 1 |
| Icacinaceae | *Icacina mannii* | Liane/Climber | | Green | 1 | 1 | 1 |
| Icacinaceae | *Leptaulus daphnoides* | Tree | Shade | Green | 0 | 1 | 1 |
| Icacinaceae | *Pyrenancantha acuminata* | Liane/Climber | Shade | Green | 0 | 0 | 1 |
| Icacinaceae | *Pyrenancantha cordicula* | Liane/Climber | Shade | Green | 1 | 0 | 0 |
| Icacinaceae | *Pyrenancantha klaineana* | Liane/Climber | | Blue | 1 | 0 | 0 |
| Icacinaceae | *Pyrenancantha vogeliana* | Liane/Climber | Shade | Green | 1 | 0 | 0 |
| Icacinaceae | *Rhaphiostylis beninensis* | Liane/Climber | Shade | Green | 0 | 0 | 1 |
| Icacinaceae | *Rhaphiostylis cordifolia* | Liane/Climber | Shade | Blue | 1 | 1 | 0 |
| Icacinaceae | *Rhaphiostylis ferruginea* | Liane/Climber | Shade | Green | 1 | 1 | 1 |
| Icacinaceae | *Rhaphiostylis preussii* | Liane/Climber | Shade | Green | 1 | 1 | 1 |
| Icacinaceae | *Stachyanthus occidentalis* | Liane/Climber | | Blue | 0 | 0 | 1 |
| Irvingiaceae | *Irvingia gabonensis* | Tree | NPLD | Green | 1 | 1 | 1 |
| Irvingiaceae | *Irvingia robur* | Tree | | Blue | 0 | 1 | 0 |
| Irvingiaceae | *Klainedoxa gabonensis* | Tree | NPLD | Green | 1 | 1 | *1* |
| Ixonanthaceae | *Phyllocosmus africanus* | Tree | Pioneer | Green | 1 | 1 | 1 |
| Ixonanthaceae | *Phyllocosmus sessiliflorus* | Shrub | Shade | Blue | 1 | 1 | 0 |
| Lecythidaceae | *Napoleonaea vogelii* | Tree | Shade | Green | 1 | 1 | 1 |
| Lecythidaceae | *Petersianthus microcarpus* | Tree | Pioneer | Green | 1 | 1 | 1 |
| Leeaceae | *Leea guineensis* | Herb | Pioneer | Green | 1 | 1 | 1 |

| Family | Scientific Name | Habit | Guild | Star | Draw River | Boi-Tano | Krokosua |
|--------|-----------------|-------|-------|------|------------|----------|----------|
| Liliaceae | *Chlorophytum orchidastrum* | Herb | | Green | 0 | 0 | 1 |
| Linaceae | *Hugonia platysepala* | Liane/Climber | NPLD | Green | 1 | 1 | 1 |
| Loganiaceae | *Anthocleista djalonensis* | Tree | Pioneer | Green | 1 | 1 | 0 |
| Loganiaceae | *Anthocleista nobilis* | Tree | Pioneer | Green | 1 | 0 | 0 |
| Loganiaceae | *Anthocleista vogelii* | Tree | Swamp | Green | 0 | 1 | 0 |
| Loganiaceae | *Mostuea brunonis* | Shrub | Pioneer | Green | 0 | 1 | 0 |
| Loganiaceae | *Strychnos aculeata* | Liane/Climber | Pioneer | Green | 1 | 1 | 1 |
| Loganiaceae | *Strychnos afzelii* | Liane/Climber | | Green | 1 | 0 | 0 |
| Loganiaceae | *Strychnos asterantha* | Liane/Climber | | Green | 1 | 1 | 0 |
| Loganiaceae | *Strychnos camptoneura* | Liane/Climber | | Green | 1 | 1 | 0 |
| Loganiaceae | *Strychnos floribunda* | Liane/Climber | NPLD | Green | 1 | 1 | 0 |
| Loganiaceae | *Strychnos johnsonii* | Liane/Climber | NPLD | Green | 0 | 0 | 1 |
| Loganiaceae | *Strychnos malacoclados* | Liane/Climber | NPLD | Green | 1 | 0 | 0 |
| Loganiaceae | *Strychnos soubrensis* | Liane/Climber | NPLD | Green | 1 | 0 | 1 |
| Loganiaceae | *Strychnos splendens* | Liane/Climber | NPLD | Green | 0 | 0 | 1 |
| Loganiaceae | *Strychnos usambarensis* | Liane/Climber | NPLD | Green | 1 | 1 | 1 |
| Loganiaceae | *Usteria guineensis* | Liane/Climber | NPLD | Green | 1 | 1 | 0 |
| Lomariopsidaceae | *Bolbitis auriculata* | Herb/Fern | | Green | 1 | 0 | 0 |
| Lomariopsidaceae | *Lomariopsis guineensis* | Herb/Fern | | Green | 1 | 0 | 0 |
| Lomariopsidaceae | *Lomariopsis palustris* | Herb/Fern | | Blue | 1 | 0 | 0 |
| Lomariopsidaceae | *Lomariopsis rossii* | Herb/Fern | | Green | 1 | 0 | 0 |
| Malpighiaceae | *Acridocarpus alternifolius* | Liane/Climber | NPLD | Green | 1 | 1 | 1 |
| Malpighiaceae | *Acridocarpus smeathmannii* | Liane/Climber | NPLD | Green | 1 | 1 | 1 |
| Malpighiaceae | *Heteropteris leons* | Liane/Climber | Pioneer | Blue | 1 | 0 | 0 |
| Maranthaceae | *Hypselodelphys poggeana* | Shrub | Pioneer | Green | 1 | 1 | 1 |
| Maranthaceae | *Hypselodelphys violacea* | Shrub | Pioneer | Green | 1 | 1 | 1 |
| Maranthaceae | *Maranthochloa purpurea* | Herb | Swamp | Green | 0 | 0 | 1 |
| Maranthaceae | *Marantochloa congensis* | Herb | Pioneer | Green | 1 | 0 | 1 |
| Maranthaceae | *Marantochloa leucantha* | Herb | Pioneer | Green | 0 | 1 | 1 |
| Maranthaceae | *Megaphrynium macrostachyum* | Herb | Pioneer | Green | 1 | 1 | 1 |
| Maranthaceae | *Sarcophrynium brachystachys* | Herb | Shade | Green | 1 | 1 | 1 |
| Maranthaceae | *Thaumatococcus danielii* | Herb | Pioneer | Red | 1 | 1 | 1 |
| Medusandraceae | *Soyauxia grandifolia* | Tree | Shade | Gold | 1 | 1 | 0 |
| Medusandraceae | *Soyauxia velutina* | Tree | Shade | Green | 0 | 1 | 0 |
| Melastomataceae | *Memecylon afzelii* | Tree | Shade | Green | 1 | 0 | 1 |
| Melastomataceae | *Memecylon aylmeri* | Tree | Shade | Blue | 1 | 1 | 0 |
| Melastomataceae | *Memecylon normandii* | Tree | Shade | Green | 1 | 0 | 0 |
| Melastomataceae | *Spathandra barteri* | Tree | Shade | Gold | 0 | 1 | 0 |
| Melastomataceae | *Tristemma coronatum* | Herb | Pioneer | Green | 1 | 0 | 0 |
| Melastomataceae | *Warneckea cinnamonoides* | Shrub | Shade | Green | 1 | 0 | 0 |

| Family | Scientific Name | Habit | Guild | Star | Draw River | Boi-Tano | Krokosua |
|--------|-----------------|-------|-------|------|------------|----------|----------|
| Melastomataceae | *Warneckea fasciculare* | Shrub | Shade | Gold | 0 | 1 | 0 |
| Melastomataceae | *Warneckea guineense* | Shrub | Shade | Green | 1 | 1 | 0 |
| Meliaceae | *Carapa procera* | Tree | Shade | Green | 1 | 1 | 1 |
| Meliaceae | *Entandrophragma angolense* | Tree | NPLD | Red | 1 | 0 | 1 |
| Meliaceae | *Entandrophragma candollei* | Tree | NPLD | Scarlet | 0 | 1 | 1 |
| Meliaceae | *Entandrophragma cylindricum* | Tree | NPLD | Scarlet | 1 | 1 | 1 |
| Meliaceae | *Entandrophragma utile* | Tree | NPLD | Scarlet | 0 | 0 | 1 |
| Meliaceae | *Guarea cedrata* | Tree | Shade | Pink | 1 | 1 | 1 |
| Meliaceae | *Guarea thompsonii* | Tree | Shade | Pink | 1 | 1 | 1 |
| Meliaceae | *Khaya ivorensis* | Tree | NPLD | Scarlet | 1 | 1 | 1 |
| Meliaceae | *Lovoa trichilioides* | Tree | NPLD | Red | 1 | 1 | 1 |
| Meliaceae | *Melia azadirach* | Tree | | Green | 0 | 1 | 0 |
| Meliaceae | *Trichilia monadelpha* | Tree | NPLD | Green | 1 | 1 | 1 |
| Meliaceae | *Trichilia preureana* | Tree | NPLD | Green | 1 | 1 | 1 |
| Meliaceae | *Trichilia tessmannii* | Tree | NPLD | Green | 1 | 1 | 1 |
| Meliaceae | *Turraeanthus africanus* | Tree | Shade | Pink | 1 | 1 | 1 |
| Melianthaceae | *Bersama abyssinica* | Tree | Pioneer | Green | 1 | 0 | 0 |
| Menispermaceae | *Albertisia scandens* | Liane/Climber | Shade | Gold | 1 | 1 | 0 |
| Menispermaceae | *Kolobopetalum leonense* | Liane/Climber | | Gold | 1 | 0 | 0 |
| Menispermaceae | *Penianthus partulinervis* | Shrub | | Green | 1 | 1 | 1 |
| Menispermaceae | *Rhigiocarya racemifera* | Liane/Climber | | Green | 0 | 0 | 1 |
| Menispermaceae | *Sphenocentrum jollyanum* | Shrub | Shade | Green | 0 | 1 | 1 |
| Menispermaceae | *Tiliacora leonensis* | Liane/Climber | NPLD | Green | 1 | 1 | 1 |
| Menispermaceae | *Triclisia patens* | Liane/Climber | | Green | 1 | 1 | 1 |
| Mimosaceae | *Acacia kamerunensis* | Liane/Climber | NPLD | Green | 1 | 0 | 1 |
| Mimosaceae | *Adenopodia scelerata* | Liane/Climber | Pioneer | Green | 1 | 0 | 0 |
| Mimosaceae | *Albizia adianthifolia* | Tree | NPLD | Green | 1 | 0 | 1 |
| Mimosaceae | *Albizia zygia* | Tree | NPLD | Green | 1 | 0 | 1 |
| Mimosaceae | *Aubrevillea platycarpa* | Tree | NPLD | Green | 0 | 1 | 0 |
| Mimosaceae | *Calpocalyx brevibracteatus* | Tree | Shade | Green | 1 | 1 | 1 |
| Mimosaceae | *Cylicodiscus gabunensis* | Tree | Shade | Blue | 0 | 1 | 1 |
| Mimosaceae | *Entada rheedei* | Liane/Climber | | Green | 0 | 1 | 0 |
| Mimosaceae | *Newtonia aubrevillei* | Tree | Shade | Gold | 1 | 1 | 1 |
| Mimosaceae | *Newtonia duparquetiana* | Tree | Shade | Blue | 1 | 1 | 0 |
| Mimosaceae | *Parkia bicolor* | Tree | NPLD | Green | 1 | 1 | 1 |
| Mimosaceae | *Pentaclethra macrophylla* | Tree | NPLD | Green | 1 | 1 | 1 |
| Mimosaceae | *Piptadeniastrum africanum* | Tree | NPLD | Pink | 1 | 1 | 1 |
| Mimosaceae | *Samanea dinklaigei* | Tree | | Blue | 1 | 1 | 0 |
| **Mimosaceae** | ***Tetrapleura chevalieri*** | **Tree** | | **Black** | **0** | **1** | **0** |
| Mimosaceae | *Xylia evansii* | Tree | NPLD | Blue | 1 | 1 | 1 |

| Family | Scientific Name | Habit | Guild | Star | Draw River | Boi-Tano | Krokosua |
|--------|-----------------|-------|-------|------|------------|----------|----------|
| Moraceae | *Antiaris toxicaria* | Tree | NPLD | Red | 1 | 1 | 1 |
| Moraceae | *Cecropia peltata* | Tree | Pioneer | Green | 1 | 1 | 1 |
| Moraceae | *Dorstenia turbinata* | Shrub | | Gold | 1 | 1 | 0 |
| Moraceae | *Ficus ardisioides* ssp. *camptoneura* | Tree/Strangler | | Blue | 0 | 1 | 0 |
| Moraceae | *Ficus craterostoma* | Tree/Strangler | | Green | 1 | 0 | 1 |
| Moraceae | *Ficus cyathistipula* | Tree | | Green | 1 | 0 | 0 |
| Moraceae | *Ficus elasticoides* | Tree/Strangler | | Blue | 1 | 0 | 0 |
| Moraceae | *Ficus exasperata* | Tree | Pioneer | Green | 0 | 0 | 1 |
| Moraceae | *Ficus kamerunensis* | Tree/Strangler | | Green | 0 | 1 | 0 |
| Moraceae | *Ficus sagittifolia* | Tree/Strangler | | Green | 0 | 0 | 1 |
| Moraceae | *Ficus sur* | Tree | Pioneer | Green | 1 | 0 | 1 |
| Moraceae | *Milicia excelsa* | Tree | Pioneer | Scarlet | 1 | 1 | 1 |
| Moraceae | *Milicia regia* | Tree | Pioneer | Scarlet | 1 | 0 | 0 |
| Moraceae | *Morus mesozygia* | Tree | | Red | 0 | 0 | 1 |
| Moraceae | *Musanga cecropioides* | Tree | Pioneer | Green | 1 | 1 | 1 |
| Moraceae | *Myrianthus arboreus* | Tree | Shade | Green | 1 | 1 | 1 |
| Moraceae | *Myrianthus libericus* | Tree | Shade | Green | 1 | 1 | 1 |
| Moraceae | *Treculia africana* | Tree | NPLD | Green | 1 | 1 | 1 |
| Moraceae | *Trilepesium madagascariense* | Tree | NPLD | Green | 1 | 1 | 1 |
| Myristicaceae | *Coelocaryon sphaerocarpum* | Tree | NPLD | Gold | 1 | 1 | 1 |
| Myristicaceae | *Pycnanthus angolensis* | Tree | NPLD | Green | 1 | 1 | 1 |
| **Myristicaceae** | ***Pycnanthus dinklaigei*** | **Liane/Climber** | **NPLD** | **Black** | **1** | **1** | **0** |
| Myrtaceae | *Eugenia calophylloides* | Shrub | Shade | Gold | 0 | 1 | 0 |
| Ochnaceae | ***Campylospermum amplectens*** | **Shrub** | **Shade** | **Black** | **1** | **1** | **0** |
| Ochnaceae | *Campylospermum laxiflorum* | Shrub | Shade | Gold | 0 | 1 | 0 |
| Ochnaceae | *Lophira alata* | Tree | Pioneer | Red | 1 | 1 | 0 |
| Ochnaceae | *Rhabdophyllum affine* ssp. *affine* | Shrub | | Green | 1 | 1 | 1 |
| Ochnaceae | *Rhabdophyllum calophyllum* | Shrub | | Green | 1 | 1 | 0 |
| Olacaceae | *Aptandra zenkeri* | Tree | Shade | Green | 1 | 0 | 1 |
| Olacaceae | *Coula edulis* | Tree | Shade | Blue | 1 | 1 | 0 |
| Olacaceae | *Heisteria parviflora* | Shrub | Shade | Green | 1 | 1 | 1 |
| Olacaceae | *Octoknema borealis* | Tree | Shade | Green | 1 | 1 | 1 |
| Olacaceae | *Olax gambecola* | Shrub | Shade | Green | 1 | 0 | 1 |
| Olacaceae | *Olax subscorpioidea* | Tree | Shade | Green | 0 | 0 | 1 |
| Olacaceae | *Ongokea gore* | Tree | NPLD | Green | 0 | 0 | 1 |
| Olacaceae | *Ptycopetallum anceps* | Shrub | Shade | Green | 1 | 1 | 0 |
| Olacaceae | *Strombosia glauscescens* | Tree | Shade | Green | 1 | 1 | 1 |
| Oleaceae | *Jasminum pauciflorum* | Shrub | NPLD | Green | 0 | 1 | 1 |
| Orchidaceae | *Eulophia guineensis* | Herb | | Green | 0 | 0 | 1 |

| Family | Scientific Name | Habit | Guild | Star | Draw River | Boi-Tano | Krokosua |
|--------|-----------------|-------|-------|------|------------|----------|----------|
| Palmae | *Calamus deeratus* | Liane/Climber | Swamp | Pink | 0 | 1 | 1 |
| Palmae | *Elaeis guineensis* | Tree | Pioneer | Pink | 1 | 1 | 1 |
| Palmae | *Eremospata hookeri* | Liane/Climber | | Pink | 1 | 1 | 1 |
| Palmae | *Eremospata macrocarpa* | Liane/Climber | | Pink | 1 | 1 | 1 |
| Palmae | *Laccosperma laeve* | Liane/Climber | NPLD | Blue | 0 | 1 | 0 |
| Palmae | *Laccosperma opacum* | Liane/Climber | NPLD | Pink | 1 | 1 | 1 |
| Palmae | *Laccosperma secundiflorum* | Liane/Climber | | Pink | 0 | 1 | 0 |
| Palmae | *Raphia hookeri* | Tree | Swamp | Green | 1 | 1 | 1 |
| Palmae | *Raphia palma-pinus* | Tree | Swamp | Gold | 1 | 1 | 0 |
| **Palmae** | ***Sclerosperma mannii*** | **Herb** | **Swamp** | **Black** | **0** | **1** | **0** |
| Pandaceae | *Panda oleosa* | Tree | Shade | Green | 1 | 1 | 1 |
| Papilionaceae | *Aganope gabonica* | Liane/Climber | Shade | Gold | 1 | 1 | 0 |
| Papilionaceae | *Aganope leucobotrya* | Liane/Climber | NPLD | Green | 1 | 1 | 0 |
| Papilionaceae | *Airyantha schweinfurthii* | Liane/Climber | NPLD | Green | 1 | 1 | 0 |
| Papilionaceae | *Angylocalyx oligophyllus* | Shrub | Shade | Green | 1 | 1 | 0 |
| Papilionaceae | *Baphia nitida* | Tree | Shade | Green | 1 | 1 | 1 |
| Papilionaceae | *Baphia pubescens* | Tree | Pioneer | Green | 1 | 1 | 1 |
| Papilionaceae | *Bowringia mildbraedii* | Liane/Climber | | Green | 1 | 1 | 0 |
| Papilionaceae | *Centrosema pubescens* | Liane/Climber | Pioneer | Green | 0 | 0 | 1 |
| Papilionaceae | *Dalbergia afzeliana* | Liane/Climber | NPLD | Green | 1 | 0 | 1 |
| Papilionaceae | *Dalbergia oblongifolia* | Liane/Climber | NPLD | Green | 1 | 1 | 0 |
| Papilionaceae | *Dioclea hexandra* | Liane/Climber | Pioneer | Green | 0 | 0 | 1 |
| Papilionaceae | *Leptoderris brachyptera* | Liane/Climber | | Green | 0 | 1 | 0 |
| Papilionaceae | *Leptoderris fasciculata* | Liane/Climber | | Green | 0 | 1 | 0 |
| Papilionaceae | *Leptoderris micrantha* | Liane/Climber | | Gold | 0 | 1 | 1 |
| **Papilionaceae** | ***Leptoderris miegei*** | **Liane/Climber** | **NPLD** | **Black** | **1** | **1** | **0** |
| Papilionaceae | *Lonchocarpus sericeus* | Tree | NPLD | Green | 0 | 0 | 1 |
| Papilionaceae | *Millettia barteri* | Liane/Climber | Swamp | Green | 0 | 0 | 1 |
| Papilionaceae | *Millettia chrysophylla* | Liane/Climber | NPLD | Green | 1 | 1 | 1 |
| Papilionaceae | *Millettia griffoniana* | Tree | Shade | Green | 1 | 1 | 0 |
| Papilionaceae | *Millettia lucens* | Liane/Climber | NPLD | Gold | 0 | 1 | 0 |
| Papilionaceae | *Millettia zechiana* | Liane/Climber | Pioneer | Green | 0 | 0 | 1 |
| Papilionaceae | *Mucuna flagellipes* | Liane/Climber | Swamp | Green | 1 | 0 | 1 |
| Papilionaceae | *Ostryocarpus riparius* | Liane/Climber | NPLD | Green | 1 | 0 | 0 |
| Papilionaceae | *Platysepalum hirsutum* | Liane/Climber | NPLD | Green | 1 | 0 | 1 |
| Papilionaceae | *Pterocarpus santalinoides* | Tree | Swamp | Green | 0 | 1 | 1 |
| Passifloraceae | *Adenia cissampeloides* | Liane/Climber | Pioneer | Green | 0 | 0 | 1 |
| Passifloraceae | *Adenia gracilis* | Liane/Climber | Pioneer | Green | 1 | 0 | 0 |
| Passifloraceae | *Adenia mannii* | Liane/Climber | Pioneer | Green | 0 | 0 | 1 |
| Passifloraceae | *Adenia rumicifolia* | Liane/Climber | Pioneer | Green | 1 | 0 | 0 |
| Passifloraceae | *Androsiphonia adenostegia* | Shrub | Shade | Blue | 1 | 1 | 0 |

| Family | Scientific Name | Habit | Guild | Star | Draw River | Boi-Tano | Krokosua |
|--------|-----------------|-------|-------|------|------------|----------|----------|
| Passifloraceae | Smeathmannia pubescens | Tree | Shade | Green | 1 | 1 | 0 |
| Phytolaccaceae | Parquetina nigrescens | Liane/Climber | Pioneer | Green | 0 | 0 | 1 |
| Piperaceae | Piper guineense | Liane/Climber | Shade | Green | 1 | 1 | 1 |
| Piperaceae | Piper umbellatum | Herb | Pioneer | Green | 0 | 0 | 1 |
| Polygalaceae | Atroxima liberica | Liane/Climber | | Green | 0 | 1 | 0 |
| Polygalaceae | Carpolobia lutea | Shrub | Shade | Green | 1 | 1 | 1 |
| Polygonaceae | Afrobrunnichia erecta | Liane/Climber | | Green | 1 | 1 | 0 |
| Polypodiaceae | Phymatodes scolopendria | Herb/Fern | | Green | 1 | 0 | 0 |
| Polypodiaceae | Platycerium stemaria | Herb/Fern | | Green | 1 | 1 | 0 |
| Rhamnaceae | Gouania longpetala | Liane/Climber | | Green | 1 | 0 | 0 |
| Rhamnaceae | Lasiodiscus fasciculiflorus | Tree | | Blue | 1 | 0 | 1 |
| Rhamnaceae | Ventilago africana | Liane/Climber | | Green | 1 | 1 | 1 |
| Rhizophoraceae | Anisophyllea meniaudii | Tree | NPLD | Gold | 1 | 1 | 0 |
| Rhizophoraceae | Anopyxis klaineana | Tree | NPLD | Red | 0 | 1 | 0 |
| **Rhizophoraceae** | ***Cassipourea biotou*** | **Tree** | | **Black** | **0** | **0** | **1** |
| Rubiaceae | Aidia genipiflora | Tree | Shade | Green | 1 | 1 | 1 |
| **Rubiaceae** | *Argocoffeffeopsis afzelii* | Shrub | **Shade** | Gold | 1 | 0 | 0 |
| Rubiaceae | Aulacocalyx jasminiflora | Tree | Shade | Green | 1 | 1 | 1 |
| Rubiaceae | Bertiera bracteolata | Liane/Climber | | Green | 1 | 0 | 0 |
| Rubiaceae | Bertiera breviflora | Shrub | | Green | 1 | 0 | 0 |
| Rubiaceae | Bertiera racemosa | Tree | Pioneer | Green | 1 | 0 | 0 |
| Rubiaceae | Chassalia afzelii | Shrub | | Green | 0 | 0 | 1 |
| Rubiaceae | Chassalia kolly | Shrub | Pioneer | Green | 0 | 0 | 1 |
| Rubiaceae | Chazaliella sciadephora | Shrub | Shade | Green | 1 | 1 | 1 |
| Rubiaceae | Coffea canephora | Tree | Shade | Pink | 1 | 0 | 0 |
| Rubiaceae | Coffea liberica | Tree | | | 0 | 1 | 0 |
| **Rubiaceae** | ***Coffea togoensis*** | **Tree** | **Shade** | **Black** | **0** | **0** | **1** |
| Rubiaceae | Corynanthe pachyceras | Tree | NPLD | Green | 1 | 1 | 1 |
| Rubiaceae | Craterispermum caudatum | Tree | Shade | Green | 1 | 1 | 1 |
| Rubiaceae | Cremaspora triflora | Liane/Climber | Shade | Green | 1 | 0 | 1 |
| Rubiaceae | Cuviera acutiflora | Tree | Pioneer | Blue | 1 | 1 | 0 |
| Rubiaceae | Didymosalpinx abbeokutae | Shrub | Shade | Blue | 1 | 1 | 1 |
| Rubiaceae | Euclinia longiflora | Shrub | | Green | 1 | 1 | 1 |
| Rubiaceae | Geophila afzelii | Shrub | Shade | Green | 1 | 1 | 0 |
| Rubiaceae | Geophila obovallata | Herb | | Green | 1 | 0 | 1 |
| Rubiaceae | Hallea ledermannii | Tree | Swamp | Red | 1 | 0 | 0 |
| Rubiaceae | Hymenocoleus hirsutus | Herb | | Green | 1 | 0 | 0 |
| Rubiaceae | Hymenocoleus libericus | Herb | | Green | 0 | 0 | 1 |
| **Rubiaceae** | ***Hymenocoleus multinervis*** | **Herb** | | **Black** | **1** | **0** | **0** |
| Rubiaceae | Ixora nigerica ssp. occidentalis | Shrub | | Green | 1 | 1 | 1 |

A biological assessment of the terrestrial ecosystems of the Draw River, Boi-Tano,
Tano Nimiri and Krokosua Hills forest reserves, southwestern Ghana

91

| Family | Scientific Name | Habit | Guild | Star | Draw River | Boi-Tano | Krokosua |
|--------|-----------------|-------|-------|------|------------|----------|----------|
| Rubiaceae | *Keetia hispida* | Shrub | Pioneer | Green | 1 | 1 | 1 |
| Rubiaceae | *Keetia rubens* | Shrub | | Gold | 1 | 1 | 0 |
| Rubiaceae | *Keetia venosa* | Shrub | Pioneer | Green | 1 | 0 | 0 |
| Rubiaceae | *Lasianthus batangensis* | Herb | Pioneer | Green | 0 | 1 | 0 |
| Rubiaceae | *Massularia acuminata* | Tree | Shade | Green | 1 | 1 | 1 |
| Rubiaceae | *Morinda lucida* | Tree | Pioneer | Green | 0 | 0 | 1 |
| Rubiaceae | *Mussaenda tristigmatica* | Liane/Climber | Pioneer | Blue | 1 | 0 | 0 |
| Rubiaceae | *Nauclea diderrichii* | Tree | Pioneer | Scarlet | 1 | 1 | 1 |
| Rubiaceae | *Nichallea soyauxii* | Liane/Climber | | Green | 1 | 1 | 0 |
| Rubiaceae | *Oxyanthus formosus* | Shrub | Shade | Green | 1 | 1 | 1 |
| Rubiaceae | *Oxyanthus pallidus* | Shrub | Shade | Green | 1 | 0 | 1 |
| Rubiaceae | *Oxyanthus speciosus* | Shrub | Shade | Green | 0 | 0 | 1 |
| Rubiaceae | *Oxyanthus subpunctatus* | Shrub | | Green | 1 | 1 | 0 |
| Rubiaceae | *Oxyanthus unilocularis* | Shrub | Shade | Green | 0 | 0 | 1 |
| Rubiaceae | *Pauridiantha sylvicola* | Tree | | Green | 1 | 1 | 0 |
| Rubiaceae | *Pausinystalia lane-poolei* | Tree | Shade | Gold | 0 | 1 | 0 |
| Rubiaceae | *Pavetta corymbosa* | Shrub | | Green | 0 | 0 | 1 |
| Rubiaceae | *Pavetta ixorifolia* | Shrub | NPLD | Green | 1 | 1 | 1 |
| Rubiaceae | *Pavetta owariensis* | Shrub | | Blue | 1 | 0 | 1 |
| Rubiaceae | *Psilanthus ebracteolatus* | Shrub | | Green | 0 | 0 | 1 |
| Rubiaceae | *Psilanthus mannii* | Shrub | | Green | 0 | 0 | 1 |
| Rubiaceae | *Psychotria albicaulis* | Shrub | Shade | Gold | 1 | 0 | 0 |
| **Rubiaceae** | ***Psychotria ankasensis*** | **Shrub** | **Shade** | **Black** | **1** | **0** | **0** |
| Rubiaceae | *Psychotria biaurita* | Shrub | Shade | Gold | 1 | 0 | 0 |
| Rubiaceae | *Psychotria brassii* | Shrub | Shade | Blue | 0 | 0 | 1 |
| Rubiaceae | *Psychotria calceata* | Shrub | Shade | Gold | 1 | 1 | 1 |
| Rubiaceae | *Psychotria elongato-sepala* | Shrub | Shade | Green | 0 | 1 | 0 |
| Rubiaceae | *Psychotria gabonica* | Shrub | Shade | Green | 0 | 1 | 1 |
| Rubiaceae | *Psychotria humilis* var. *major* | Shrub | Shade | Gold | 0 | 1 | 0 |
| Rubiaceae | *Psychotria ivorensis* | Shrub | Shade | Gold | 1 | 1 | 0 |
| Rubiaceae | *Psychotria kitsonii* | Shrub | Shade | Gold | 1 | 0 | 0 |
| Rubiaceae | *Psychotria liberica* | Shrub | Shade | Gold | 1 | 1 | 0 |
| Rubiaceae | *Psychotria mangenotii* | Shrub | Shade | Gold | 1 | 0 | 0 |
| Rubiaceae | *Psychotria peduncularis* | Shrub | Shade | Green | 1 | 1 | 1 |
| Rubiaceae | *Psychotria rufipilis* | Shrub | Shade | Gold | 0 | 1 | 0 |
| Rubiaceae | *Psychotria schweinfurthii* | Shrub | Shade | Green | 0 | 0 | 1 |
| Rubiaceae | *Psychotria subobliqua* | Shrub | Shade | Green | 0 | 1 | 1 |
| Rubiaceae | *Psydrax horizontale* | Shrub | Pioneer | Green | 1 | 0 | 0 |
| Rubiaceae | *Psydrax manense* | Tree | Pioneer | Green | 1 | 0 | 0 |
| Rubiaceae | *Psydrax parviflora* | Tree | Pioneer | Green | 0 | 0 | 1 |
| Rubiaceae | *Rothmania hispida* | Tree | Shade | Green | 1 | 1 | 1 |

| Family | Scientific Name | Habit | Guild | Star | Draw River | Boi-Tano | Krokosua |
|---|---|---|---|---|---|---|---|
| Rubiaceae | *Rothmania longiflora* | Tree | Shade | Green | 0 | 1 | 1 |
| Rubiaceae | *Rothmania whitfieldii* | Tree | Shade | Green | 1 | 1 | 1 |
| Rubiaceae | *Sarcocephalus pobeguinii* | Tree | Swamp | Green | 0 | 1 | 0 |
| **Rubiaceae** | ***Schumanniophyton problematicum*** | **Tree** | **Shade** | **Black** | **0** | **1** | **0** |
| Rubiaceae | *Sericanthe chevalieri* | Tree | | Blue | 1 | 0 | 0 |
| **Rubiaceae** | ***Sericanthe toupetou*** | **Tree** | **Shade** | **Black** | **1** | **0** | **0** |
| Rubiaceae | *Sherbournia calycina* | Liane/Climber | | Blue | 1 | 0 | 0 |
| Rubiaceae | *Vangueriella orthacantha* | Liane/Climber | | Blue | 1 | 0 | 0 |
| Rubiaceae | *Vangueriella vanguerioides* | Liane/Climber | | Blue | 1 | 0 | 0 |
| Rutaceae | *Afraegle paniculata* | Tree | non-forest | Blue | 0 | 0 | 1 |
| Rutaceae | *Citropsis articulata* | Tree | Shade | Blue | 0 | 0 | 1 |
| Rutaceae | *Vepris soyauxii* | Tree | | Green | 1 | 1 | 0 |
| Rutaceae | *Zanthoxylon gillotii* | Tree | Pioneer | Green | 1 | 1 | 1 |
| Santalaceae | *Okoubaka aubrevillei* | Tree | NPLD | Gold | 0 | 1 | 0 |
| Sapindaceae | *Allophylus africanus* | Tree | Pioneer | Green | 0 | 1 | 0 |
| Sapindaceae | *Allophylus talbotii* | Liane/Climber | | Gold | 1 | 0 | 0 |
| **Sapindaceae** | ***Apodiscus chevalieri*** | **Tree** | | **Black** | **0** | **1** | **0** |
| Sapindaceae | *Aporrhiza urophylla* | Tree | Shade | Green | 1 | 1 | 0 |
| Sapindaceae | *Blighia sapida* | Tree | NPLD | Green | 1 | 1 | 1 |
| Sapindaceae | *Blighia unijugata* | Tree | Shade | Green | 1 | 0 | 1 |
| Sapindaceae | *Blighia welwitschii* | Tree | NPLD | Green | 1 | 1 | 1 |
| Sapindaceae | *Chytranthus atroviolaceus* | Tree | Shade | Green | 0 | 1 | 1 |
| Sapindaceae | *Chytranthus carneus* | Tree | Shade | Green | 1 | 1 | 1 |
| Sapindaceae | *Chytranthus cauliflorus* | Shrub | Shade | Blue | 1 | 1 | 0 |
| Sapindaceae | *Chytranthus macrobotrys* | Tree | Swamp | Green | 1 | 0 | 1 |
| **Sapindaceae** | ***Chytranthus verecundus*** | **Shrub** | **Shade** | **Black** | **0** | **1** | **0** |
| Sapindaceae | *Deinbollia pinnata* | Shrub | NPLD | Green | 0 | 1 | 1 |
| Sapindaceae | *Deinbollia saligna* | Shrub | Swamp | Gold | 1 | 0 | 0 |
| Sapindaceae | *Erioceolum pungens* | Tree | Swamp | Green | 1 | 1 | 0 |
| Sapindaceae | **Erioceolum racemosum** | Tree | Swamp | Blue | 1 | 1 | 0 |
| Sapindaceae | *Lecaniodiscus cupanioides* | Tree | Shade | Green | 0 | 0 | 1 |
| Sapindaceae | *Lepisanthes senegalensis* | Tree | Shade | Green | 0 | 0 | 1 |
| Sapindaceae | *Majidea fosteri* | Tree | NPLD | Green | 0 | 0 | 1 |
| Sapindaceae | *Pancovia bijuga* | Tree | Shade | Green | 0 | 0 | 1 |
| Sapindaceae | *Pancovia pedicellaris* | Tree | Shade | Green | 1 | 1 | 1 |
| Sapindaceae | *Pancovia sesiliflora* | Shrub | Shade | Green | 0 | 1 | 1 |
| Sapindaceae | *Paullinia pinnata* | Liane/Climber | Pioneer | Green | 0 | 1 | 1 |
| **Sapindaceae** | ***Placodiscus bancoensis*** | **Tree** | **NPLD** | **Black** | **1** | **1** | **0** |
| Sapindaceae | *Placodiscus boya* | Tree | Shade | Gold | 1 | 0 | 0 |
| Sapindaceae | *Placodiscus oblongifolius* | Tree | Shade | Gold | 1 | 1 | 0 |

| Family | Scientific Name | Habit | Guild | Star | Draw River | Boi-Tano | Krokosua |
|---|---|---|---|---|---|---|---|
| Sapotaceae | *Chrysophyllum africanum var aubrevillei* | Tree | Shade | Blue | 0 | 1 | 0 |
| Sapotaceae | *Chrysophyllum albidum* | Tree | Shade | Pink | 0 | 0 | 1 |
| Sapotaceae | *Chrysophyllum giganteum* | Tree | Shade | Pink | 0 | 1 | 1 |
| Sapotaceae | *Chrysophyllum pruniforme* | Tree | Shade | Green | 1 | 1 | 0 |
| Sapotaceae | *Chrysophyllum purpulchrum* | Tree | NPLD | Green | 0 | 0 | 1 |
| Sapotaceae | *Chrysophyllum subnudum* | Tree | Shade | Green | 1 | 1 | 1 |
| Sapotaceae | *Chrysophyllum ubangiense* | Tree | | Blue | 0 | 1 | 0 |
| Sapotaceae | *Chrysophyllum welwitschii* | Shrub | Shade | Green | 1 | 0 | 1 |
| Sapotaceae | *Delpydora gracillis* | Shrub | Shade | Gold | 1 | 1 | 0 |
| Sapotaceae | *Manilkara obovata* | Tree | Shade | Blue | 1 | 1 | 0 |
| Sapotaceae | *Omphalocarpum ahia* | Tree | Swamp | Blue | 1 | 1 | 0 |
| Sapotaceae | *Omphalocarpum procerum* | Tree | NPLD | Blue | 1 | 1 | 0 |
| Sapotaceae | *Omphalocarpum procerum* | Tree | NPLD | Blue | 1 | 1 | 0 |
| Sapotaceae | *Pouteria alnifolia* | Tree | Pioneer | Green | 0 | 0 | 1 |
| Sapotaceae | *Pouteria altissima* | Tree | NPLD | Red | 0 | 0 | 1 |
| Sapotaceae | *Pouteria aningeri* | Tree | NPLD | Pink | 0 | 0 | 1 |
| Sapotaceae | *Synsepalum afzelii* | Tree | Shade | Green | 1 | 1 | 1 |
| Sapotaceae | *Tieghemella heckelii* | Tree | NPLD | Scarlet | 1 | 1 | 1 |
| Scytopetalaceae | *Rhaptopetalum beguei* | Tree | Swamp | Gold | 1 | 0 | 0 |
| Scytopetalaceae | *Scytopetalum tieghemii* | Tree | Shade | Blue | 1 | 1 | 0 |
| Simaroubaceae | *Hannoa klaineana* | Tree | Pioneer | Green | 1 | 1 | 1 |
| Simaroubaceae | *Pierreodendron kerstingii* | Tree | Shade | Gold | 1 | 0 | 0 |
| Smilacaceae | *Smilax anceps* | Liane/Climber | Pioneer | Green | 0 | 0 | 1 |
| Solanaceae | *Solanum erianthum* | Shrub | Pioneer | Green | 0 | 0 | 1 |
| Solanaceae | *Solanum torvum* | Shrub | Pioneer | Green | 0 | 0 | 1 |
| Sterculiaceae | *Byttneria catalpifolia africana* | Liane/Climber | | Blue | 1 | 1 | 0 |
| Sterculiaceae | *Cola caricifolia* | Tree | Shade | Green | 1 | 1 | 1 |
| Sterculiaceae | *Cola chlamydantha* | Tree | Shade | Blue | 1 | 1 | 0 |
| Sterculiaceae | *Cola digitata* | Tree | Pioneer | Green | 1 | 1 | 0 |
| Sterculiaceae | *Cola gigantea* | Tree | NPLD | Green | 1 | 1 | 1 |
| Sterculiaceae | *Cola heterophylla* | Shrub | Shade | Green | 1 | 1 | 0 |
| Sterculiaceae | *Cola lateritea* | Tree | Shade | Green | 1 | 1 | 1 |
| Sterculiaceae | *Cola millenii* | Tree | NPLD | Green | 0 | 0 | 1 |
| Sterculiaceae | *Cola nitida* | Tree | Shade | Pink | 1 | 1 | 1 |
| Sterculiaceae | *Cola reticulata* | Tree | Shade | Green | 1 | 0 | 0 |
| **Sterculiaceae** | ***Cola umbratilis*** | **Tree** | **Shade** | **Black** | **1** | **1** | **0** |
| Sterculiaceae | *Heritiera utilis* | Tree | NPLD | Red | 1 | 1 | 0 |
| Sterculiaceae | *Hildegardia barteri* | Tree | NPLD | Blue | 0 | 0 | 1 |
| Sterculiaceae | *Mansonia altissima* | Tree | NPLD | Pink | 0 | 0 | 1 |
| Sterculiaceae | *Nesogordonia papaverifera* | Tree | Shade | Pink | 0 | 0 | 1 |

| Family | Scientific Name | Habit | Guild | Star | Draw River | Boi-Tano | Krokosua |
|---|---|---|---|---|---|---|---|
| Sterculiaceae | *Pterygota bequaertii* | Tree | NPLD | Red | 1 | 1 | 0 |
| Sterculiaceae | *Pterygota macrocarpa* | Tree | NPLD | Red | 0 | 0 | 1 |
| Sterculiaceae | *Scaphopetalum amoenum* | Tree | Shade | Blue | 1 | 1 | 0 |
| Sterculiaceae | *Sterculia oblonga* | Tree | NPLD | Green | 1 | 1 | 1 |
| Sterculiaceae | *Sterculia rhinopetala* | Tree | NPLD | Pink | 0 | 0 | 1 |
| Sterculiaceae | *Sterculia tragacantha* | Tree | Pioneer | Green | 0 | 0 | 1 |
| Sterculiaceae | *Theobroma cacao* | Tree | non-forest | | 0 | 0 | 1 |
| Sterculiaceae | *Triplochiton scleroxylon* | Tree | Pioneer | Scarlet | 0 | 0 | 1 |
| Thymeleaceae | *Craterosiphon scandens* | Liane/Climber | | Green | 1 | 0 | 0 |
| Thymeleaceae | *Dicranolepis persei* | Shrub | Shade | Green | 1 | 1 | 1 |
| Tiliaceae | *Christiana africana* | Tree | | Green | 0 | 0 | 1 |
| Tiliaceae | *Desplatsia chrysochlamys* | Tree | Shade | Green | 1 | 1 | 1 |
| Tiliaceae | *Desplatsia dewevrei* | Tree | Shade | Green | 1 | 0 | 0 |
| Tiliaceae | *Desplatsia subericarpa* | Tree | Shade | Green | 0 | 1 | 1 |
| Tiliaceae | *Duboscia viridiflora* | Tree | NPLD | Green | 0 | 0 | 1 |
| Tiliaceae | *Glyphaea brevis* | Tree | Shade | Green | 0 | 1 | 1 |
| Tiliaceae | *Grewia carpinifolia* | Liane/Climber | NPLD | Green | 1 | 0 | 0 |
| Tiliaceae | *Grewia hookerana* | Liane/Climber | NPLD | Green | 1 | 1 | 0 |
| Tiliaceae | *Grewia malacocarpa* | Liane/Climber | NPLD | Green | 1 | 1 | 0 |
| Tiliaceae | *Grewia mollis* | Liane/Climber | Pioneer | Green | 0 | 1 | 0 |
| Ulmaceae | *Celtis adolfi-friderici* | Tree | Pioneer | Green | 0 | 0 | 1 |
| Ulmaceae | *Celtis mildbraedii* | Tree | Shade | Green | 0 | 0 | 1 |
| Ulmaceae | *Celtis phillipensis* | Tree | | Green | 0 | 0 | 1 |
| Ulmaceae | *Celtis zenkeri* | Tree | NPLD | Green | 0 | 0 | 1 |
| Ulmaceae | *Holoptelea grandis* | Tree | NPLD | Green | 0 | 0 | 1 |
| Ulmaceae | *Trema orientalis* | Tree | Pioneer | Green | 0 | 1 | 1 |
| Urticaceae | *Boehmeria platyphylla* | Herb | Pioneer | Green | 1 | 0 | 0 |
| Urticaceae | *Urera robusta* | Liane/Climber | Pioneer | Gold | 1 | 0 | 1 |
| Verbenaceae | *Clerodendrum silvanum* | Liane/Climber | Pioneer | Blue | 1 | 0 | 0 |
| Verbenaceae | *Clerodendrum silvanum* var. *bucholzii* | Liane/Climber | Pioneer | Green | 1 | 0 | 0 |
| Verbenaceae | *Clerodendrum capitatum* | Liane/Climber | Pioneer | Green | 1 | 0 | 0 |
| Verbenaceae | *Vitex ferruginea* | Tree | NPLD | Green | 0 | 0 | 1 |
| Verbenaceae | *Vitex grandifolia* | Tree | NPLD | Green | 1 | 0 | 1 |
| Violaceae | *Decorsella paradoxa* | Tree | Shade | Gold | 1 | 1 | 0 |
| Violaceae | *Rinorea angustifolia* | Tree | Shade | Green | 0 | 0 | 1 |
| Violaceae | *Rinorea brachypetala* | Shrub | Shade | Blue | 1 | 0 | 0 |
| Violaceae | *Rinorea breviracemosa* | Shrub | Shade | Blue | 1 | 0 | 0 |
| Violaceae | *Rinorea ilicifolia* | Shrub | Shade | Green | 0 | 0 | 1 |
| Violaceae | *Rinorea oblongifolia* | Tree | Shade | Green | 1 | 1 | 1 |
| Violaceae | *Rinorea prasina* | Shrub | Shade | Blue | 1 | 1 | 0 |

| Family | Scientific Name | Habit | Guild | Star | Draw River | Boi-Tano | Krokosua |
|---|---|---|---|---|---|---|---|
| Violaceae | *Rinorea subintegrifolia* | Shrub | Shade | Green | 1 | 0 | 0 |
| Violaceae | *Rinorea welwitschii* | Tree | Shade | Green | 1 | 0 | 0 |
| Violaceae | *Rinorea yaundensis* | Tree | Shade | Green | 0 | 0 | 1 |
| Vitaceae | *Cissus aralioides* | Liane/Climber | | Green | 1 | 1 | 1 |
| Vitaceae | *Cissus producta* | Liane/Climber | | Green | 1 | 0 | 1 |
| Zingiberaceae | *Costus deistelii* | Herb | Pioneer | Green | 1 | 0 | 0 |
| Zingiberaceae | *Costus dubius* | Herb | Pioneer | Green | 1 | 0 | 0 |
| Zingiberaceae | *Costus engleranus* | Herb | Swamp | Green | 1 | 1 | 1 |
| Zingiberaceae | *Renealmia battenbergiana* | Herb | Shade | Gold | 1 | 1 | 0 |
| | | | | | 470 | 399 | 378 |

# Appendix 2

## Checklist of the Butterflies of Ghana with a list of those collected during the RAP survey in western Ghana, including those previously known from the adjacent National Parks of Ankasa and Bia

*Torben Larsen*

This checklist is an updated version of the one published by Emmel and Larsen (1997). It includes all butterflies known from Ghana. Since original publication of the list, there have been extensive revisions of a number of large genera (*Ornipholidotos, Torbenia, Epitola, Lachnocnema, Euphaedra, Euptera* and others), some newly described species, and a number of species new to Ghana collected in the field.

There have also been many changes in taxonomy and nomenclature. The families and subfamilies are revised more or less on the basis of the GlobIS Project where a number of prominent systematists are gradually reaching a consensus on the higher classification of butterflies worldwide. This has been much aided by recent advances in dna-based cladistic analysis. The most radical change is probably the removal of the *Argynnis*-group of genera from the Nymphalinae to the Heliconiinae, where the former subfamily Acraeinae is also now placed as one of the tribes.

The genera, species, and subspecies are as they will appear in my book: *Butterflies of West Africa – origins, natural history, diversity, conservation.* At the time of writing the manuscript is under final editing.

In November/December 2000 I did a butterfly survey of Ankasa and Bia National Parks, including a brief visit to Krokosua Hills (Larsen 2001). About 120 species were collected at Krokosua, twenty of which were not found during the present survey. Since Ankasa and Bia are very close to the RAP localities, the data from this survey are included here for comparative purposes.

### Abbreviations

The checklist attempts to incorporate as much information as possible in a limited space. This has only been possible through the use of symbols and abbreviations as listed below:

**Numbers:** The numbering refers to the manuscript of *Butterflies of West Africa – origins, natural history, diversity, conservation.* Missing numbers refer to butterflies not found in Ghana, mainly Nigerian species and a limited number of endemics of the extreme west of West Africa (Liberian subregion). There are also changes in number sequencing because of revisions in higher level taxonomy.

**Species:** Each butterfly is listed with the name of the Genus, species, subspecies, author and year of description. The names are as up-to-date as possible and generally agreed as valid amongst the limited number of researchers dealing with the butterflies of tropical Africa.

**Endemics:** Given the purposes of the RAP, endemic butterflies are flagged by the following superscripts after the year of description:
GHE – endemic to the Ghana subregion of West Africa (from the Volta Region to Central Côte d'Ivoire - 25 species
WAE – endemic to Africa west of the Dahomey Gap - 61 species
WNE – endemic to West Africa, but extending to Nigeria (and in some cases the extreme west of Cameroon in similar habitats) - 63 species

**VOL** – species known in West Africa only from Volta Region (some endemic to Volta Region, others extending to Nigeria, and a few further east than Nigeria) – 10 species

## Habitat choice:

ALF    widely distributed in forests without clear preference

WEF    centered on the wettest forest (wet evergreen forest)

MEF    centered on the most forests (moist evergreen and semi-deciduous)

DRF    centered on drier forests (coastal and bordering Guinea savannah)

GUI    centered on the Guinea savannah ('middle belt')

SUD    centered on the Sudan savannahs of the extreme north of Ghana

UBQ    found everywhere except in wetter primary forests

SPE    centered on special habitats (rocky areas, grasslands, swamps) in forest and savannah# indicates that a species has been recorded from within an area protected by Ghana Wildlife + collected during expeditions by the *Butterflies of West Africa Project*

## Ankasa, Bia, or Krokosua:

**AN**  known with certainty from Ankasa National Park
**BI**  known with certainty from Bia National Park
**an**  occurs in Ankasa with 80% probability
**bi**  occurs in Bia with 80% probability
**??**  may occur in the locality
**~**   collected in the Krokosua Hills during the 2000 Bia Survey
**--**  Ghana butterflies that could not occur in the locality for geographical or ecological reasons (e.g. species not occurring west of the Volta or extreme Sudan savannah species, marked as such).

**Status**: This column given a rough-and-ready summary of the relative rarity of the butterflies in Ghana. Some of these may be common in other parts of Africa. There is no objective measurement for this. Some common butterfly may be unaccountably absent from a locality for a period of time. Some rare butterflies may be common in a given place on a single date (*Acraea kraka kibi* has been found only about five times in West Africa, but on three or four of these there were veritable population explosions). Rarity can be due to many reasons: Genuine rarity, very low population densities, habits that make them difficult to collect except by special methods, very narrow habitat choices, etc). There are also geographical differences; some species might not be rare in Sierra Leone but very rare in Ghana. Nonetheless, rare species are mainly found in 'good' habitats and there is a surprising degree of unanimity on the 'rarity' of species in both the literature and in my debriefings with experienced lepidopterists.

1    very common – species found on almost any thorough visit to a suitable habitat
2    common – species found on 75% of thorough visits to a suitable habitat
3    not rare – species found regularly (25-75%} of thorough visits to a suitable habitat
4    rare – species found on less than 10-20% of thorough visits to a suitable habitat
5    very rare – species often never found during the series of thorough visits to a suitable habitat (no more than 10-20 records from Africa)

## Collected in RAP survey localities:

**DRR** -   Draw River Forest Reserve
**BOT** -   Boi-Tano Forest Reserve
**KRO** -   Krokosua Hills (**KRX** species caught at Krokosua in 2000 but not during the RAP)
**RAP** -   all species collected during the RAP expedition

Emmel, T. and T.B. Larsen. 1997. Butterfly diversity in Ghana, West Africa. Tropical Lepidoptera, 8 (supplement 3):1-13.

Larsen, T.B. 2001. The butterflies of Ankasa/Nini-Suhien and Bia protected area systems in western Ghana. Protected Areas Development Programme. UGL/Ghana Wildlife Department.

| FAMILY | Species name | Habitat choice | collected # | collected + | Ankasa | Bia | Krokosua | Status | DRR | BOT | KRO | RAP total |
|---|---|---|---|---|---|---|---|---|---|---|---|---|
| **PAPILIONIDAE** | | | | | | | | | | | | |
| **Papilioninae** | | | | | | | | | | | | |
| Papilionini | | | | | | | | | | | | |
| 1 | *Papilio antimachus antimachus* Drury, 1782 | WEF | | | ?? | Bi | | 4 | | | | |
| 2 | *Papilio zalmoxis* Hewitson, 1864 | WEF | # | | an | BI | | 4 | | | | |
| 4 | *Papilio dardanus dardanus* Brown, 1776 | ALF | # | + | AN | BI | | 3 | DRR | BOT | KRO | RAP |
| 5 | *Papilio phorcas phorcas* Cramer, 1775 | ALF | # | + | ?? | ?? | | 4 | | | | |
| 7 | *Papilio horribilis* Butler, 1874[WAE] | WEF | # | + | AN | bi | | 3 | DRR | | | RAP |
| 9 | *Pap chrapkowskoides nurettini* Koçak, 1983 | ALF | # | + | AN | BI | | 2 | | | | |
| 10 | *Pap sosia sosia* Rothschild & Jordan, 1903 | ALF | # | + | AN | BI | | 3 | | | | |
| 11 | *Papilio nireus nireus* Linné, 1758 | ALF | # | + | AN | BI | ~ | 2 | DRR | BOT | KRO | RAP |
| 12 | *Papilio menestheus menestheus* Drury, 1773 | WEF | # | + | AN | BI | ~ | 2 | DRR | BOT | KRO | RAP |
| 13 | *Papilio demodocus demodocus* Esper, 1798 | UBQ | # | + | AN | BI | ~ | 1 | DRR | BOT | KRO | RAP |
| 15 | *Papilio cyproeofila* Butler, 1868 | MEF | # | + | AN | BI | | 2 | DRR | BOT | | RAP |
| 16 | *Papilio zenobia* Fabricius, 1775 | MEF | # | + | AN | BI | | 3 | | | | |
| 17 | *Papilio nobicea* Suffert, 1904[VOE] | MEF | # | + | x | x | | 3 | | | | |
| 18 | *Papilio cynorta cynorta* Fabricius, 1793 | MEF | # | + | AN | BI | ~ | 3 | | | KRO | RAP |
| Leptocircini | | | | | | | | | | | | |
| 20 | *Graphium angolanus baronis* Ungemach, 1932 | GUI | # | + | x | bi | | 2 | | | | |
| 22 | *Graphium tynderaeus* Fabricius, 1793 | WEF | # | + | AN | BI | | 4 | | | | |
| 23 | *Graphium l. latreillianus* Godart, 1819 | WEF | # | + | AN | ?? | | 3 | | BOT | | RAP |
| 24 | *Graphium almansor* Honrath, 1884 | DRF | # | + | x | x | | 3 | | | | |
| 25 | *Graphium adamastor* Boisduval, 1836 | DRF | # | + | x | x | | 3 | | | | |
| 26 | *Graphium agamedes* Westwood, 1842 | DRF | | | x | x | | 4 | | | | |
| 27 | *Graphium rileyi* Berger, 1950[WAE] | WEF | | | ?? | ?? | | 4 | | | | |
| 28 | *Graphium leonidas leonidas* Fabricius, 1793 | UBQ | # | + | AN | BI | ~ | 2 | DRR | BOT | KRO | RAP |
| 29 | *Graphium illyris* Hewitson, 1873 | WEF | # | + | an | BI | | 3 | | BOT | | RAP |
| 30 | *Graphium policenes* Cramer, 1775 | ALF | # | + | AN | BI | ~ | 2 | DRR | BOT | KRO | RAP |
| 31 | *Graphium liponesco* Suffert, 1904 | WEF | # | + | AN | BI | | 3 | | | | |
| 32 | *Graphium antheus* Cramer, 1779 | ALF | # | + | an | BI | | 4 | | | | |

| FAMILY | # | Species name | Habitat choice | collected # | collected + | Ankasa | Bia | Krokosua | Status | DRR | BOT | KRO | RAP total |
|---|---|---|---|---|---|---|---|---|---|---|---|---|---|
| **PIERIDAE** | | | | | | | | | | | | | |
| **Pseudopontiinae** | | | | | | | | | | | | | |
| Coliadinae | 33 | *Pseudopontia paradoxa paradoxa* F. & F.,1869 | WEF | # | + | AN | ?? | . | 3 | | | | |
| | 34 | *Catopsilia florella* Fabricius, 1793 | UBQ | # | + | AN | BI | ' | 1 | DRR | BOT | KRO | RAP |
| | 36 | *Eurema senegalensis* Boisduval, 1836 | MEF | # | + | AN | BI | ' | 2 | DRR | BOT | KRO | RAP |
| | 37 | *Eurema becabe solifera* Butler, 1875 | UBQ | # | + | AN | BI | ' | 1 | DRR | BOT | KRO | RAP |
| | 38 | *Eurema floricola leonis* Butler, 1886 | UBQ | # | + | AN | BI | | 3 | | | | |
| | 40 | *Eurema desjardinsii regularis* Butler, 1876 | UBQ | # | + | an | bi | | 3 | | | | |
| | 41 | *Eurema brigitta brigitta* Stoll, 1870 | GUI | # | + | AN | BI | ' | 2 | | | KRO | RAP |
| **Pierinae** | 42 | *Pinacopteryx eriphia tritogenia* Klug, 1829 | SUD | | | x | x | | 2 | | | | |
| | 43 | *Nepheronia argia argia* Fabricius, 1775 | ALF | # | + | AN | BI | ' | 2 | DRR | BOT | KRO | RAP |
| | 44 | *Nepheronia thalassina thalassina* Bsd., 1836 | ALF | # | + | AN | BI | ' | 2 | | | KRO | RAP |
| | 45 | *Nepheronia pharis pharis* Boisduval, 1836 | MEF | # | + | AN | BI | ' | 3 | | | KRO | RAP |
| | 52 | *Colotis vesta amelia* Lucas, 1852 | SUD | | + | x | x | | 4 | | | | |
| | 55 | *Colotis celimene sudanicus* Aurivillius, 1905 | SUD | | | x | x | | 4 | | | | |
| | 56 | *Colotis ione* Godart, 1819 | SUD | # | | x | x | | 3 | | | | |
| | 58 | *Colotis danae eupompe* Klug, 1829 | SUD | | | x | x | | 3 | | | | |
| | 59 | *Colotis aurora evarne* Klug, 1829 | SUD | | + | x | x | | 3 | | | | |
| | 60 | *Colotis antevippe antevippe* Boisduval, 1836 | SUD | # | + | x | x | | 3 | | | | |
| | 61 | *Colotis euippe euippe* Linné, 1758 | UBQ | # | + | AN | BI | ' | 2 | DRR | | KRO | RAP |
| | 63 | *Colotis evagore antigone* Boisduval, 1836 | SUD | # | + | x | bi | | 2 | | | | |
| | 66 | *Belenois aurota* Fabricius, 1793 | SUD | # | + | an | BI | | 1 | | | | |
| | 67 | *Belenois creona creona* Cramer, 1776 | SUD | # | + | an | bi | | 2 | | | | |
| | 68 | *Belenois gidica* Godart, 1819 | GUI | # | + | x | x | | 3 | | | | |
| | 70 | *Belenois subeida frobeniusi* Strand, 1909 | SUD | # | + | x | x | | 3 | | | | |
| | 71 | *Belenois calypso calypso* Drury, 1773 | ALF | # | + | AN | BI | ' | 1 | DRR | | KRO | RAP |
| | 72 | *Belenois theora theora* Doubleday, 1846 | ALF | # | + | AN | BI | ' | 2 | | | KRO | RAP |
| | 74 | *Belenois hedyle rhena* Doubleday, 1846[WNE] | DRF | # | + | x | bi | | 3 | | | | |
| | 76 | *Dixeia doxo doxo* Godart, 1819 | SUD | | | x | bi | | 3 | | | | |
| | 77 | *Dixeia orbona orbona* Geyer, 1832 | GUI | | | x | x | | 3 | | | | |

| FAMILY / No. | Species name | Habitat choice | collected | | Ankasa / Bia / Krokosua | | | Status | RAP collection localities | | | RAP total |
|---|---|---|---|---|---|---|---|---|---|---|---|---|
| 78 | Dixeia cebron Westwood, 1871 | DRF | # | + | x | bi | | 3 | | | KRO | RAP |
| 79 | Dixeia capricornus capricornus West., 1871 | DRF | # | + | ?? | bi | | 3 | | | KRO | RAP |
| 82 | Appias sylvia sylvia Fabricius, 1775 | ALF | # | + | AN | BI | ʼ | 2 | DRR | BOT | KRO | RAP |
| 83 | Appias phaola phaola Doubleday 1847 | MEF | # | + | AN | ?? | ʼ | 3 | | | KRO | RAP |
| 84 | Appias sabina sabina Felder & Felder, 1865 | ALF | # | + | AN | BI | ʼ | 2 | DRR | BOT | KRO | RAP |
| 85 | Appias epaphia epaphia Boisduval, 1833 | UBQ | # | + | AN | bi | ʼ | 2 | | | KRO | RAP |
| 86 | Leptosia alcesta alcesta Stoll, 1784 | ALF | # | + | AN | BI | ʼ | 1 | DRR | BOT | KRO | RAP |
| 88 | Leptosia hybrida hybrida Bernardi, 1952 | MEF | # | + | AN | BI | ʼ | 2 | DRR | BOT | KRO | RAP |
| 89 | Leptosia medusa Cramer, 1777 | MEF | # | + | AN | BI | ʼ | 2 | DRR | BOT | KRO | RAP |
| 90 | Leptosia marginea Mabille, 1890 | MEF | # | + | AN | BI | | 3 | | | | |
| 91 | Leptosia wigginsi pseudalcesta Bern., 1964 | ALF | # | + | AN | BI | | 3 | | | KRO | RAP |
| 93 | Mylothris chloris chloris Fabricius, 1775 | UBQ | # | + | an | BI | | 1 | | BOT | KRO | RAP |
| 98 | Mylothris dimidiata Aurivillius, 1898[WAE] | WEF | # | + | AN | ?? | | 3 | | | | |
| 101 | Mylothris aburi Larsen & Collins, 2003[WNE] | GUI | # | + | x | x | | 3 | | | | |
| 104 | Mylothris poppea Cramer, 1777[WAE] | MEF | # | + | AN | BI | | 3 | | | | |
| 105 | Mylothris spica Möschler, 1884[GHE] | MEF | # | + | an | BI | | 3 | | | | |
| 107 | Mylothris rhodope Fabricius, 1775 | ALF | # | + | AN | BI | | 2 | | | KRO | RAP |
| 108 | Mylothris jaopura Karsch, 1893 | ALF | # | + | AN | BI | | 2 | | | | |
| 109 | Mylothris schumanni schumanni Suffert, 1904 | MEF | # | + | AN | BI | | 3 | | | | |
| 110 | Mylothris atewa Berger, 1980[GHE] | WEF | # | + | x | x | | 3 | | | | |
| **LYCAENIDAE** | | | | | | | | | | | | |
| **Miletinae** | | | | | | | | | | | | |
| Liphyrini | | | | | | | | | | | | |
| 319 | Euliphyra hewitsoni Aurivillius, 1898 | MEF | # | + | AN | BI | | 4 | | | | |
| 320 | Euliphyra mirifica Holland, 1890 | MEF | # | | x | x | | 4 | | | | |
| 321 | Euliphyra leucyania Hewitson, 1874 | WEF | | | an | bi | | 4 | | | | |
| 322 | Aslauga ernesti Karsch, 1895[VOL] | DRF | | | x | x | | 5 | | | | |
| 323 | Aslauga marginalis Kirby, 1890 | MEF | # | + | an | BI | | 3 | | | | |
| 326 | Aslauga lamborni Bethune-Baker, 1914 | WEF | # | + | an | BI | | 4 | | | | |
| 327 | Aslauga ashanti Libert, in press[GHE] | WEF | | | ?? | ?? | | 5 | | | | |

A biological assessment of the terrestrial ecosystems of the Draw River, Boi-Tano, Tano Nimiri and Krokosua Hills forest reserves, southwestern Ghana

101

| FAMILY | | Species name | Habitat choice | collected | | Ankasa | Bia / Krokosua | Status | RAP collection localities | RAP total |
|---|---|---|---|---|---|---|---|---|---|---|
| Miletinae | 330 | Aslauga imitans Libert, 1994[VOL] | MEF | | | x | x | 4 | | |
| | 335 | Megalopalpus zymna Hewitson, 1852 | ALF | # | + | AN | BI | 2 | DRR | RAP |
| | 337 | Megalopalpus metaleucus Karsch, 1893 | MEF | # | + | AN | BI | 3 | | |
| Spalgini | 338 | Spalgis lemolea Druce, 1890 | DRF | # | + | an | BI | 3 | | |
| Lachncnemini | 339 | Lachnocnema vuattouxi Libert, 1996 | DRF | # | + | an | BI | 3 | | |
| | 341 | Lachnocnema emperanus Snellen, 1872 | DRF | # | + | AN | bi | 3 | | |
| | 343 | Lachnocnema disrupta Talbot, 1935 | MEF | | | ?? | ?? | 4 | | |
| | 344 | Lachnocnema r. reutlingeri Holland, 1892 | MEF | | | an | bi | 4 | | |
| | 345 | Lachnocnema luna Druce, 1910 | WEF | | | x | x | 4 | | |
| | 347 | Lachnocnema albimacula Libert, 1996 | WEF | | | ?? | ?? | 4 | | |
| **Lipteninae** Pentilini | 116 | Ptelina carnuta Hewitson, 1873 | WEF | # | + | AN | BI | 3 | | |
| | 117 | Pentila pauli pauli Staudinger, 1888 | DRF | # | + | an | bi | 3 | KRO | RAP |
| | 121 | Pentila petreia Hewitson, 1874[WNE] | WEF | # | + | AN | BI | 2 | KRO | RAP |
| | 126 | Pentila picena Hewitson, 1874[WNE] | WEF | # | + | AN | BI | 3 | | |
| | 129 | Pentila phidia Hewitson, 1874[GHE] | MEF | # | + | x | BI | 3 | KRX | |
| | 131 | Pentila hewitsonii hewitsonii S. & K. 1887 | MEF | # | + | an | BI | 3 | KRO | RAP |
| | 133 | Telipna acraea acraea Hewitson, 1851 | WEF | # | + | AN | bi | 3 | | |
| | 134 | Telipna semirufa Grose-Smith & Kirby, 1889[WAE] | MEF | # | + | AN | BI | 3 | | |
| | 135 | Telipna maesseni Stempffer, 1970[GHN] | WEF | # | + | x | x | 3 | | |
| | 142 | Ornipholidotos camerunensis Stempffer, 1964 | WEF | | + | an | bi | 4 | | |
| | 143 | Ornipholidotos kirbyi Aurivillius, 1895[GHN] | WEF | | | an | bi | 4 | | |
| | 144 | Ornipholidotos nigeriae Stempffer, 1964 | WEF | # | + | an | BI | 4 | | |
| | 144a | Ornipholidotos irwini Collins & Larsen, 1998 | WEF | | | ?? | x | 5 | | |
| | 145 | Ornipholidotos onitshae Stempffer, 1962 | WEF | # | + | an | BI | 4 | | |
| | 147 | Ornipholidotos tiassale Stempffer, 1969[WNE] | MEF | # | + | AN | BI | 3 | | |
| | 148 | Ornipholidotos nympha Holland, 1890[WNE] | WEF | # | + | an | BI | 4 | | |

| FAMILY | No. | Species name | Habitat choice | collected (#) | collected (+) | Ankasa (AN) | Bia (BI) | Krokosua | Status | DRR | BOT | KRO | RAP total |
|---|---|---|---|---|---|---|---|---|---|---|---|---|---|
| | 149 | *Torbenia larseni* Stempffer, 1969 | WEF | # | + | AN | ?? | | 4 | | | KRO | RAP |
| Mimacraeini | 149a | *Torbenia wojtusiaki* Libert, 2001 [WNE] | WEF | | | an | x | | 5 | | | | |
| | 150 | *Mimacraea neurata* Holland, 1895 | WEF | # | + | an | bi | | 4 | | | | |
| | 152 | *Mimacraea darwinia* Butler, 1872 [WAE] | WEF | # | + | an | BI | | 3 | | | | |
| | 152a | *Mimacraea maesseni* Libert, 2000 [GHN] | WEF | # | + | x | x | | 3 | | | | |
| | 153 | *Mimeresia libentina* Hewitson, 1866 | ALF | # | + | AN | BI | ʼ | 2 | DRR | | | |
| | 154 | *Mimeresia moyambina* Bethune-Baker, 1904 [WAE] | WEF | | | an | ?? | | 5 | | | | |
| | 155 | *Mimeresia debora catori* Bethune-Baker, 1904 | WEF | | | ?? | ?? | | 4 | | | | |
| | 156 | *Mimeresia semirufa* Grose-Smith, 1902 [WAE] | WEF | # | + | AN | BI | ʼ | 4 | | | KRX | |
| | 159 | *Mimeresia cellularis* Kirby, 1890 | WEF | | | an | x | | 4 | | | | |
| | 160 | *Mimeresia issia* Stempffer, 1969 [WAE] | WEF | # | + | AN | x | | 4 | | | | |
| Liptenini | 161 | *Pseuderesia eleaza eleaza* Hewitson, 1873 | WEF | # | + | an | bi | | 3 | | | | |
| | 162 | *Eresiomera bicolor* Sm. & Kirby, 1887 [WNE] | WEF | # | + | AN | BI | | 3 | DRR | BOT | KRO | RAP |
| | 163 | *Eresiomera isca occidentalis* C. & L., 1998 | WEF | # | + | an | BI | | 4 | | | | |
| | 164 | *Eresiomera jacksoni* Stempffer, 1969 [GHE] | WEF | | | ?? | ?? | | 5 | | | | |
| | 166 | *Eresiomera petersi* St.& Bennett, 1956 [GHE] | MEF | # | + | AN | BI | | 4 | | | | |
| | 168 | *Citrinophila marginalis* Kirby, 1887 [WNE] | MEF | # | + | AN | BI | | 2 | | | | |
| | 169 | *Citrinophila similis* Kirby, 1887 [WNE] | MEF | # | + | AN | BI | ʼ | 2 | DRR | | KRO | RAP |
| | 171 | *Citrinophila erastus erastus* Hewitson, 1866 | WEF | # | + | an | BI | | 4 | DRR | | KRO | RAP |
| | 173 | *Eresina maesseni* Stempffer, 1956 [WNE] | MEF | | + | an | bi | | 4 | | | KRO ? | RAP |
| | 176 | *Eresina pseudofusca* Stempffer, 1961 [WNE] | MEF | # | | an | BI | | 4 | | | | |
| | 179 | *Eresina saundersi* Stempffer, 1956 | MEF | | | an | bi | | 4 | | | | |
| | 180 | *Eresina rougeoti* Stempffer, 1956 | MEF | | | an | bi | | 4 | | | | |
| | 181 | *Eresina theodori* Stempffer, 1956 [WNE] | MEF | # | + | an | BI | | 3 | | | | |
| | 182 | *Argyrocheila undifera undifera* Staud., 1892 | WEF | # | + | an | BI | | 4 | | | | |
| | 185 | *Liptena submacula liberiana* SBM, 1974 [WNE] | MEF | # | + | AN | BI | | 3 | | | | |
| | 187 | *Liptena simplicia* Möschler, 1887 [WNE] | MEF | # | + | AN | BI | | 2 | | | | |
| | 191 | *Liptena tiassale* Stempffer, 1969 [GHE] | MEF | | | x | bi | | 4 | | | | |
| | 193 | *Liptena albicans* Cator, 1904 [WNE] | WEF | | | an | bi | | 4 | | | | |

A biological assessment of the terrestrial ecosystems of the Draw River, Boi-Tano, Tano Nimiri and Krokosua Hills forest reserves, southwestern Ghana

103

| FAMILY | No. | Species name | Habitat choice | collected | | Ankasa | Bia | Krokosua | Status | RAP collection localities | | | RAP total |
|---|---|---|---|---|---|---|---|---|---|---|---|---|---|
| | 194 | *Liptena alluaudi* Mabille, 1890[WNE] | WEF | # | + | an | BI | | 3 | | | KRO | RAP |
| | 194 | *Liptena fatima* Kirby, 1890 | DRF | | | x | ?? | | 5 | | | | |
| | 195 | *Liptena pearmani* St., Bennett & May, 1974[GHN] | WEF | | + | x | x | | 4 | | | | |
| | 199 | *Liptena septistrigata* Bethune-Baker, 1903 | DRF | | | x | bi | | 3 | | | | |
| | 200 | *Liptena evanescens* Kirby, 1887 | WEF | | | an | bi | | 4 | | | | |
| | 202 | *Liptena xanthostola coomassiensis* H.-S., 1933 | WEF | | + | AN | BI | | 4 | | | | |
| | 203 | *Liptena bia* sp nov | MEF | # | + | x | BI | | 5 | | | | |
| | 204 | *Liptena rochei* Stempffer, 1951[WNE] | DRF | | | x | bi | | 4 | | | | |
| | 205 | *Liptena flavicans oniens* Talbot, 1935 | MEF | | | x | bi | | 4 | | | | |
| | 207 | *Liptena similis* Kirby, 1890[WNE] | WEF | # | + | an | BI | | 4 | | | | |
| | 209 | *Liptena helena* Druce, 1888[WNE] | WEF | # | + | AN | bi | | 3 | | | | |
| | 210 | *Liptena catalina* S. & K., 1890 | WEF | # | + | AN | bi | | 3 | DRR | | | RAP |
| | 213 | *Kakumia otlauga* Grose-Smith & Kirby, 1890[WNE] | WEF | # | + | AN | BI | | 3 | | | | |
| | 216 | *Falcuna leonensis* St. & Bennett, 1963[WAE] | WEF | # | + | AN | BI | ˎ | 2 | DRR | | KRO | RAP |
| | 219 | *Falcuna campinus* Holland, 1890 | WEF | # | + | AN | BI | | 3 | | | | |
| | 221 | *Tetrarhanis symplocus* Clench, 1965[WNE] | MEF | # | + | AN | BI | | 2 | | | KRO | RAP |
| | 222 | *Tetrarhanis baralingam* Larsen, 1998[WAE] | WEF | # | + | AN | BI | | 4 | DRR | | | RAP |
| | 227 | *Tetrarhanis stempfferi stempfferi* B, 1954 | WEF | | + | x | ?? | | 4 | | | | |
| | 231 | *Larinopoda aspidos* Druce, 1890[GHN] | MEF | # | + | x | x | | 3 | | | | |
| | 232 | *Larinopoda eurema* Plötz, 1880[WAE] | MEF | # | + | AN | BI | ˎ | 2 | DRR | BOT | KRO | RAP |
| | 233 | *Micropentila adelgitha* Hewitson, 1874 | WEF | # | + | AN | BI | ˎ | 2 | | | KRX | |
| | 234 | *Micropentila adelgunda* Staudinger, 1892 | MEF | # | + | an | BI | | 5 | | | | |
| | 235 | *Micropentila dorothea* Bethune-Baker, 1903 | MEF | | + | an | bi | | 3 | | | | |
| | 237 | *Micropentila brunnea brunnea* Kirby, 1887 | WEF | # | + | AN | bi | | 4 | | | | |
| | 237a | *Micropentila cf katerae* St. & Bennett, 1965 | WEF | | | an | bi | | 4 | | | | |
| | 242 | *Micropentila mamfe* Larsen, 1986[WAE] | WEF | | + | an | bi | | 5 | | | | |
| Epitolini | | | | | | | | | | | | | |
| | 245 | *Iridana rougeoti* Stempffer, 1964 | ALF | | + | an | bi | | 4 | | | | |
| | 246 | *Iridana ghanana* Stempffer, 1964[VOL] | ALF | | | x | x | | 4 | | | | |
| | 247 | *Iridana exquisita* Grose-Smith, 1898 | MEF | # | | AN | bi | | 5 | | | | |
| | 248 | *Iridana nigeriana* Stempffer, 1964[WNE] | ALF | | + | an | bi | | 4 | | | | |

| FAMILY | | Species name | Habitat choice | collected | | Ankasa / Bia / Krokosua | | Status | RAP collection localities | RAP total |
|---|---|---|---|---|---|---|---|---|---|---|
| | 249 | *Hewitsonia boisduvalii* Hewitson, 1869 | WEF | # | + | AN | BI | 3 | | |
| | 250 | *Hewitsonia occidentalis* Bouyer, 1997[WNE] | MEF | | + | an | bi | 3 | | |
| | 252 | *Hewitsonia inexpectata* Bouyer, 1997 | MEF | # | + | AN | BI | 4 | | |
| | 255 | *Cerautola crowleyi crowleyi* Sharpe, 1890 | MEF | | + | an | bi | 3 | | |
| | 257 | *Cerautola ceraunia* Hewitson, 1879 | MEF | # | + | an | bi | 3 | DRR | RAP |
| | 260 | *Epitola posthumus* Fabricius, 1793 | MEF | # | + | an | BI | 3 | | |
| | 261 | *Epitola uranoides occidentalis* Libert, 1999 | WEF | | + | an | bi | 4 | | |
| | 262 | *Epitola urania* Kirby, 1887 | WEF | | | ?? | ?? | 4 | | |
| | 263 | *Cephetola cephena cephena* Hewitson, 1873 | MEF | # | + | an | bi | 3 | | |
| | 265 | *Cephetola pinodes pinodes* Druce, 1890 | MEF | | | ?? | bi | 4 | | |
| | 266 | *Cephetola subcoerulea* Roche, 1954 | MEF | # | + | ?? | ?? | 4 | | |
| | 267 | *Cephetola nigra* Bethune-Baker, 1903 | MEF | | + | an | bi | 3 | | |
| | 268 | *Cephetola mercedes ivoriensis* Jack., 1967 | MEF | | + | an | bi | 4 | | |
| | 269 | *Cephetola obscura* Hawker-Smith, 1926[WNE] | MEF | # | + | an | bi | 4 | | |
| | 271 | *Cephetola sublustris* Bethune-Baker, 1904 | MEF | | | an | bi | 4 | | |
| | 272 | *Cephetola maesseni* Libert, 1999[WAE] | MEF | | | an | bi | 4 | | |
| | 273 | *Cephetola collinsi* Libert & Larsen, 1999[GHE] | MEF | | + | an | bi | 3 | | |
| | 274 | *Hypophytala hyettoides* Aurivillius, 1895 | MEF | # | + | an | bi | 3 | | |
| | 276 | *Hypophytala hyettina* Aurivillius, 1897 | MEF | | | an | bi | 4 | | |
| | 277 | *Hypophytala henleyi* Kirby, 1890 | MEF | | + | an | bi | 4 | | |
| | 278 | *Hypophytala benitensis benitensis* Hol., 1890 | WEF | # | | AN | ?? | 4 | | |
| | 280 | *Phytala elais elais* Westwood, 1851 | WEF | | | an | bi | 4 | | |
| | 281 | *Geritola gerina* Hewitson, 1878 | WEF | | + | an | bi | 4 | | |
| | 286 | *Geritola virginea* Bethune-Baker, 1904 | WEF | | + | an | bi | 4 | | |
| | 288 | *Stempfferia cercene* Hewitson, 1873 | WEF | # | + | an | bi | 4 | | |
| | 290 | *Stempfferia moyambina* Bethune-Baker, 1903[WNE] | WEF | # | + | an | bi | 3 | DRR | RAP |
| | 292 | *Stempfferia dorothea* Bethune-Baker, 1904[WAE] | WEF | # | + | an | BI | 3 | | |
| | 296 | *Stempfferia leonina* Staudinger, 1888[WAE] | WEF | # | + | an | bi | 3 | | |
| | 298 | *Stempfferia ciconia ciconia* G-K. & K., 1892 | WEF | # | + | an | bi | 3 | | |
| | 300 | *Stempfferia zelza* Hewitson, 1878[VOL] | WEF | | | x | x | 4 | | |
| | 306 | *Stempfferia michelae michelae* Libert, 1999 | ALF | | + | an | BI | 3 | | |

| FAMILY | No. | Species name | Habitat choice | collected # | collected + | Ankasa | Bia / Krokosua | Status | RAP collection localities | RAP total |
|---|---|---|---|---|---|---|---|---|---|---|
|  | 310 | *Stempfferia staudingeri* Kirby, 1890[WAE] | WEF |  |  | an | bi | 4 |  |  |
|  | 312 | *Aethiopana honorius divisa* Butler, 1901 | WEF | # | + | AN | BI | 3 |  |  |
|  | 313 | *Epitolina dispar* Kirby, 1887 | MEF | # | + | AN | BI | 2 | BOT | RAP |
|  | 314 | *Epitolina melissa* Druce, 1888 | MEF | # | + | AN | BI | 2 | DRR | RAP |
|  | 316 | *Epitolina catori catori* Beth.-Baker, 1904 | MEF | # | + | an | BI | 3 |  |  |
|  | 317 | *Neaveia lamborni lamborni* Druce, 1910 | MEF |  | + | an | bi | 4 |  |  |
| **Theclinae** |  |  |  |  |  |  |  |  |  |  |
| Amblypodiini |  |  |  |  |  |  |  |  |  |  |
|  | 378 | *Myrina silenus silenus* Fabricius, 1775 | GUI | # | + | AN | bi | 3 |  |  |
|  | 379 | *Myrina subornata* Lathy, 1903 | GUI |  |  | x | bi | 4 |  |  |
| Oxylidini |  |  |  |  |  |  |  |  |  |  |
|  | 380 | *Oxylides faunus faunus* Drury, 1773 | MEF | # | + | AN | BI | 2 | DRR, KRO | RAP |
| Aphnaeini |  |  |  |  |  |  |  |  |  |  |
|  | 350 | *Aphnaeus orcas* Drury, 1782 | MEF | # | + | AN | BI | 3 |  |  |
|  | 351 | *Aphnaeus argyrocyclus* Holland, 1890 | MEF | # | + | an | BI | 4 |  |  |
|  | 352 | *Aphnaeus asterius* Plötz, 1880 | WEF | # |  | an | bi | 4 |  |  |
|  | 353 | *Aphnaeus brahami* Lathy, 1903 | GUI |  |  | x | bi | 4 |  |  |
|  | 355 | *Aphnaeus gilloni* Stempffer, 1966 | MEF |  |  | ?? | ?? | 5 |  |  |
|  | 356 | *Apharitis nilus* Hewitson, 1865 | SUD |  |  | x | x | 4 |  |  |
|  | 357 | *Spindasis mozambica* Bertolini, 1850 | GUI | # | + | x | ?? | 3 |  |  |
|  | 358 | *Spindasis avriko* Karsch, 1893 | GUI |  |  | x | x | 4 |  |  |
|  | 359 | *Spindasis crustaria* Holland, 1890[VOL] | WEF |  |  | x | x | 4 |  |  |
|  | 360 | *Spindasis iza* Hewitson, 1865[WAE] | WEF |  | + | an | bi | 4 |  |  |
|  | 361 | *Spindasis menelas* Druce, 1907[WNE] | WEF |  |  | an | bi | 5 |  |  |
|  | 362 | *Zeritis neriene* Boisduval, 1836 | SUD |  |  | x | x | 3 |  |  |
|  | 363 | *Axiocerses amanga* Westwood, 1881 | GUI | # | + | AN | BI | 3 |  |  |
|  | 365 |  | SUD |  |  | x | x | 3 |  |  |
|  | 366 | *Lipaphnaeus leonina ivoirensis* St., 1966 | MEF | # | + | an | bi | 3 |  |  |
|  | 367 | *Lipaphnaeus aderna aderna* Plötz, 1880 | GUI |  | + | x | bi | 3 |  |  |
|  | 368a | *Pseudaletis cf agrippina*[WAE] | MEF |  |  | an | bi | 5 |  |  |
|  | 272a | *Pseudaletis subangulata* Talbot, 1935[WAE] | DRF |  |  | an | bi | 4 |  |  |

| | FAMILY | Species name | Habitat choice | collected | | Ankasa | Bia / Krokosua | Status | RAP collection localities | RAP total |
|---|---|---|---|---|---|---|---|---|---|---|
| 374 | | *Pseudaletis dardanella* Riley, 1922 | MEF | | | an | bi | 5 | | |
| 377 | | *Pseudaletis leonis* Staudinger, 1888[WNE] | MEF | # | | an | bi | 4 | | |
| | ??? Loxurini | | | | | | | | | |
| 383 | | *Dapidodigma hymen* Fabricius, 1775[WNE] | MEF | # | + | AN | BI | 3 | BOT | |
| 384 | | *Dapidodigma demeter demeter* Clench, 1961 | MEF | # | + | AN | BI | 4 | KRX | RAP |
| | Iolaini | | | | | | | | | |
| 385 | | *Iolaus eurisus eurisus* Cramer, 1779 | ALF | # | + | AN | BI | 3 | | |
| 386 | | *Iolaus alienus bicaudatus* Aurivillius, 1905 | SUD | | | x | x | 4 | | |
| 389 | | *Iolaus scintillans* Aurivillius, 1905 | SUD | | | x | x | 3 | | |
| 390 | | *Iolaus laon laon* Hewitson, 1878 | MEF | # | + | an | bi | 3 | | |
| 393 | | *Iolaus banco* Stempffer, 1966 | MEF | | | an | ?? | 3 | | |
| 400 | | *Iolaus sappirus* Druce, 1902 | MEF | # | + | AN | BI | 3 | | |
| 402 | | *Iolaus bellina bellina* Plötz, 1880 | WEF | # | + | an | bi | 4 | | |
| 406 | | *Iolaus fontainei* Stempffer, 1956 | WEF | | | x | x | 4 | | |
| 408 | | *Iolaus aethria* Karsch, 1893 | MEF | | + | an | bi | 4 | | |
| 409 | | *Iolaus farquharsoni* Bethune-Baker, 1922 | MEF | | + | ?? | ?? | 4 | | |
| 410 | | *Iolaus iasis iasis* Hewitson, 1865 | ALF | # | + | an | BI | 3 | | |
| 411 | | *Iolaus maesa* Hewitson, 1862 | MEF | | + | ?? | ?? | 4 | | |
| 413 | | *Iolaus parasilanus maeseni* St. & B., 1958 | MEF | # | + | x | x | 3 | | |
| 414 | | *Iolaus menas menas* Druce, 1890 | SUD | # | + | x | x | 3 | | |
| 416 | | *Iolaus carolinae* Collins & Larsen, 2000[GHE] | MEF | | + | ?? | ?? | 5 | | |
| 417 | | *Iolaus ismenias* Klug, 1834 | SUD | | + | x | bi | 3 | | |
| 419 | | *Iolaus iulus* Hewitson, 1869 | MEF | # | + | an | bi | 3 | | |
| 422 | | *Iolaus alcibiades* Kirby, 1871 | MEF | | + | an | bi | 4 | | |
| 423 | | *Iolaus lukabas* Druce, 1890[WNE] | MEF | | | ?? | ?? | 3 | | |
| 423a | | *Iolaus mane* Collins & Larsen, 2003[WAE] | MEF | | | x | ?? | 4 | | |
| 424 | | *Iolaus theodori* Stempffer, 1970[GHE] | MEF | | | x | x | 4 | | |
| 425 | | *Iolaus paneperata* Druce, 1890 | MEF | # | + | ?? | ?? | 3 | | |
| 426 | | *Iolaus likpe* Collins & Larsen, 2003[GHE] | MEF | | | ?? | bi | 5 | | |
| 426a | | *Iolaus kibi* sp. nov.[GHE] | WEF | | | ?? | ?? | 5 | | |
| 427 | | *Iolaus calisto* Westwood, 1851 | WEF | # | + | an | BI | 3 | | |

| FAMILY | | Species name | Habitat choice | collected | | Ankasa / Bia / Krokosua | | Status | RAP collection localities | RAP total |
|---|---|---|---|---|---|---|---|---|---|---|
| | 428 | *Iolaus laonides* Aurivillius, 1897 | MEF | | | ?? | ?? | 4 | | |
| | 429 | *Iolaus timon timon* Fabricius, 1787 | MEF | | | ?? | bi | 4 | | |
| | 430 | *Etesiolaus catori catori* Beth.-Baker, 1904 | DRF | # | + | AN | bi | 4 | | |
| | 431 | *Etesiolaus kyabobo* Larsen, 1996 | DRF | # | + | ?? | ?? | 4 | | |
| | 432 | *Stugeta marmoreus marmoreus* Butler, 1866 | GUI | | | x | x | 3 | | |
| Hypolycaenini | | | | | | | | | | |
| | 434 | *Hypolycaena philippus philippus* F. 1793 | GUI | # | + | AN | bi | 2 | DRR | RAP |
| | 435 | *Hypolycaena kadiskos* Druce, 1890 | MEF | | | ?? | ?? | 4 | | |
| | 436 | *Hypolycaena liara liara* Druce, 1890 | MEF | | + | an | bi | 4 | | |
| | 437 | *Hypolycaena lebona lebona* Hewitson, 1867 | WEF | # | + | an | BI | 3 | DRR | RAP |
| | 438 | *Hypolycaena clenchi* Larsen, 1997[WAE] | WEF | # | + | AN | BI | 4 | DRR | RAP |
| | 440 | *Hypolycaena scintillans* Stempffer, 1957[WNE] | MEF | # | + | AN | BI | 2 | KRO | RAP |
| | 441 | *Hypolycaena dubia* Aurivillius, 1895 | ALF | # | + | AN | BI | 2 | BOT | RAP |
| | 442 | *Hypolycaena kakumi* Larsen, 1997 | MEF | # | + | AN | bi | 2 | | |
| | 443 | *Hypolycaena antifaunus antifaunus* W., 1851 | MEF | # | + | an | BI | 3 | | |
| | 444 | *Hypolycaena hatita hatita* Hewitson, 1865 | MEF | # | + | AN | BI | 2 | DRR | RAP |
| | 446 | *Hypolycaena nigra* Bethune-Baker, 1914 | WEF | # | + | an | BI | 3 | | |
| Deudorigini | | | | | | | | | | |
| | 448 | *Hypomyrina nomenia* Hewitson, 1874[WNE] | DRF | | + | ?? | Bi | 4 | | |
| | 449 | *Hypomyrina nomion* Staudinger, 1891 | DRF | # | | ?? | bi | 3 | | |
| | 452 | *Kopelates virgata* Druce, 1891 | WEF | # | + | AN | ?? | 4 | | |
| | 456 | *Hypokopelates otraeda* Hewitson, 1863 | MEF | # | + | AN | bi | 3 | | |
| | 457 | *Hypokopelates dimitris* D'Abrera, 1980 | MEF | # | + | an | bi | 4 | | |
| | 458 | *Hypokopelates leonina* Bethune-Baker, 1904 | MEF | # | + | an | bi | 3 | | |
| | 460 | *Hypokopelates viridis viridis* St., 1964[WNE] | ALF | # | + | AN | BI | 3 | KRO | RAP |
| | 469 | *Hypokopelates catori* Bethune-Baker, 1903[WNE] | DRF | # | + | x | bi | 4 | | |
| | 471 | *Pilodeudorix caerulea caerulea* Druce, 1891 | GUI | # | + | ?? | BI | 3 | | |
| | 472 | *Pilodeudorix camerona camerona* Plötz, 1880 | WEF | # | + | an | BI | 3 | | |
| | 473 | *Pilodeudorix diyllus diyllus* Hewitson, 1878 | WEF | # | + | an | BI | 3 | | |
| | 474 | *Pilodeudorix zela* Hewitson, 1869 | WEF | # | + | an | BI | 4 | | |
| | 475 | *Diopetes kakumi* Larsen, 1994 | WEF | # | + | an | bi | 5 | | |

Checklist of the Butterflies of Ghana with a list of those collected during the
RAP survey in western Ghana, including those previously known from the
adjacent National Parks of Ankasa and Bia

| FAMILY | No. | Species name | Habitat choice | collected # | collected + | Ankasa | Bia | Krokosua | Status | DRR | BOT | KRO | RAP total |
|---|---|---|---|---|---|---|---|---|---|---|---|---|---|
| | 479 | *Diopetes violetta* Aurivillius, 1897 | WEF | | + | an | bi | | 4 | | | | |
| | 480 | *Diopetes aurivilliusi* Stempffer, 1954[WNE] | WEF | | + | an | bi | | 4 | | | | |
| | 481 | *Diopetes fumata* Stempffer, 1954 | WEF | | | ?? | ?? | | 5 | | | | |
| | 483 | *Deudorix odana* Druce, 1887 | ALF | # | + | an | bi | | 3 | | | | |
| | 484 | *Deudorix galathea* Swainson, 1821[WNE] | MEF | # | + | AN | bi | | 3 | | | | |
| | 485 | *Deudorix antalus* Hopffer, 1855 | GUI | # | + | an | bi | | 2 | | | | |
| | 486 | *Deudorix caliginosa* Lathy, 1903 | MEF | # | + | ?? | ?? | | 4 | | | | |
| | 487 | *Deudorix dinomenes diomedes* Jackson, 1966 | DRF | # | + | ?? | BI | | 4 | | | | |
| | 489 | *Deudorix dinochares* Grose-Smith, 1887 | GUI | # | + | ?? | ?? | | 4 | | | | |
| | 490 | *Deudorix lorisona lorisona* Hewitson, 1862 | ALF | # | + | an | BI | | 3 | | | | |
| | 491 | *Deudorix kayonza maesseni* Libert, in prep. | WEF | # | + | an | BI | | 4 | | | | |
| | 492a | *Capys vorgasi* Collins & Larsen, 2003[GHE] | DRF | | | x | x | | 5 | | | | |
| **Polyommatinae** | | | | | | | | | | | | | |
| Lycaenesthini | | | | | | | | | | | | | |
| | 493 | *Anthene rubricinctus rubricinctus* H., 1891 | MEF | # | + | AN | BI | ' | 2 | DRR | BOT | KRO | RAP |
| | 494 | *Anthene ligures* Hewitson, 1874 | MEF | # | + | an | bi | | 4 | | | | |
| | 495 | *Anthene sylvanus sylvanus* Drury, 1773 | ALF | # | + | AN | BI | | 2 | DRR | BOT | | RAP |
| | 497 | *Anthene liodes liodes* Hewitson, 1874 | ALF | # | + | AN | bi | | 3 | | BOT | | RAP |
| | 498 | *Anthene definita definita* Butler, 1899 | GUI | # | + | x | bi | | 3 | | | | |
| | 499 | *Anthene princeps princeps* Butler, 1876 | GUI | | + | x | bi | | 3 | | | | |
| | 500 | *Anthene nigropunctata* Bethune-Baker, 1910 | GUI | | + | x | ?? | | 4 | | | | |
| | 501 | *Anthene amarah* Guérin-Méneville, 1847 | SUD | # | + | x | ?? | | 2 | | | | |
| | 502 | *Anthene lunulata* Trimen, 1894 | GUI | # | + | x | ?? | | 3 | DRR | | | RAP |
| | 503 | *Anthene kikuyu* Bethune-Baker, 1910 | GUI | | + | x | x | | 4 | | | | |
| | 504 | *Anthene talboti* Stempffer, 1936 | SUD | # | + | x | x | | 4 | | | | |
| | 505 | *Anthene wilsoni* Talbot, 1935 | GUI | | + | x | x | | 4 | | | | |
| | 506 | *Anthene levis* Hewitson, 1878 | ALF | # | + | an | BI | | 3 | | | | |
| | 507 | *Anthene pungusei* manuscript name[GHE] | DRF | | | x | x | | 5 | | | | |
| | 508 | *Anthene irumu* Stempffer, 1948 | ALF | | + | an | bi | | 3 | | | | |
| | 509 | *Anthene larydas* Cramer, 1780 | ALF | # | + | AN | BI | ' | 2 | DRR | BOT | KRO | RAP |
| | 510 | *Anthene crawshayi crawshayi* Butler, 1899 | GUI | # | + | x | x | | 3 | | | | |

| FAMILY | No. | Species name | Habitat choice | collected | | Ankasa | Bia | Krokosua | Status | RAP collection localities | RAP total |
|---|---|---|---|---|---|---|---|---|---|---|---|
|  | 511 | *Anthene lachares lachares* Hewitson, 1878 | MEF | # | + | an | BI |  | 3 |  |  |
|  | 513 | *Anthene lysicles* Hewitson, 1874 | WEF | # | + | AN | bi |  | 3 |  |  |
|  | 516 | *Anthene atewa* Larsen & Collins, 1998[GHE] | WEF |  | + | AN | x |  | 4 |  |  |
|  | 518 | *Anthene radiata* Bethune-Baker, 1910[WAE] | WEF |  |  | an | x |  | 5 |  |  |
|  | 520 | *Anthene locuples* Grose-Smith, 1898 | WEF | # | + | ?? | x |  | 4 |  |  |
|  | 523 | *Anthene scintillula aurea* Bethune-Baker, 1910 | WEF | # |  | an | bi |  | 4 |  |  |
|  | 524 | *Anthene helpsi* Larsen, 1994[GHE] | WEF |  | + | ?? | ?? |  | 5 |  |  |
|  | 525 | *Anthene juba* Fabricius, 1787[WNE] | WEF | # | + | AN | bi |  | 3 |  |  |
|  | 526 | *Neurypexina lysianus* Hewitson, 1874 | WEF | # | + | AN | BI |  | 2 |  |  |
|  | 528 | *Neurellipes lusones* Hewitson, 1874 | WEF | # | + | AN | ?? |  | 4 |  |  |
|  | 529 | *Neurellipes chryseostictus* Beth.-B., 1910 | WEF | # | + | an | bi |  | 3 |  |  |
|  | 530 | *Neurellipes fulvus* Stempffer, 1962 | WEF | # | + | AN | x |  | 5 |  |  |
|  | 531 | *Neurellipes staudingeri* S. & K., 1894 | WEF |  |  | an | ?? |  | 4 |  |  |
|  | 532 | *Neurellipes gemmifera* Neave, 1910 | DRF | # | + | x | x |  | 4 |  |  |
|  | 533 | *Triclema rufoplagata* Bethune-Baker, 1910 | MEF |  |  | an | bi |  | 4 |  |  |
|  | 534 | *Triclema lucretilis lucretilis* Hew., 1874 | MEF | # | + | an | bi |  | 3 |  |  |
|  | 535 | *Triclema lamias lamias* Hewitson, 1878 | MEF | # | + | ?? | BI |  | 3 |  |  |
|  | 536 | *Triclema fasciatus* Aurivillius, 1895 | WEF | # | + | an | bi | ˎ | 3 | KRX |  |
|  | 537 | *Triclema obscura* Bethune-Baker, 1910 | WEF |  |  | an | ?? |  | 4 |  |  |
|  | 538 | *Triclema inconspicua* Druce, 1910 | WEF | # | + | an | BI | ˎ | 4 |  |  |
|  | 540 | *Triclema hades* Bethune-Baker, 1910 | MEF | # | + | ?? | bi | ˎ | 3 |  |  |
|  | 541 | *Triclema phoenicis* Karsch, 1893 | DRF | # | + | x | bi |  | 4 |  |  |
|  | 542 | *Triclema nigeriae* Aurivillius, 1905[WNE] | GUI | # | + | x | bi |  | 3 |  |  |
|  | 546 | *Capidesthes jacksoni* Stempffer, 1969[GHE] | WEF | # | + | AN | BI |  | 3 | DRR | RAP |
|  | 548 | *Capidesthes lithas* Druce, 1890 | MEF | # | + | AN | BI |  | 3 | KRX |  |
|  | 550 | *Capidesthes leonina* Bethune-Baker, 1903 | MEF | # | + | an | BI |  | 3 |  |  |
|  | 551 | *Capidesthes* sp. nov.[GHE] | MEF | # | + | ?? | ?? |  | 5 |  |  |
| Polyommatini | 552 | *Pseudonacaduba sichela sichela* Wall., 1857 | GUI | # | + | AN | BI |  | 2 | DRR  BOT  KRO | RAP |
|  | 554 | *Lampides boeticus* Linné, 1767 | UBQ | # | + | AN | bi |  | 2 |  |  |
|  | 555 | *Uranothauma falkensteini* Dewitz, 1879 | ALF | # | + | AN | BI |  | 2 | KRO | RAP |

| No. | Species name | Habitat choice | collected # | collected + | Ankasa | Bia | Krokosua | Status | RAP loc. DRR | RAP loc. BOT | RAP loc. KRO | RAP total |
|---|---|---|:-:|:-:|:-:|:-:|:-:|:-:|:-:|:-:|:-:|:-:|
| 561 | *Phlyaria cyara stactalla* Karsch, 1895 | ALF | # | + | an | BI |  | 2 |  |  |  |  |
| 562 | *Cacyreus lingeus* Stoll, 1782 | DRF | # | + | an | BI | ı | 2 |  |  | KRO | RAP |
| 564 | *Cacyreus audeoudi* Stempffer, 1936 | WEF | # |  | an | bi |  | 4 |  |  |  |  |
| 565 | *Leptotes pirithous* Linné, 1767 | UBQ | # | + | an | BI | ı | 2 |  |  | KRX |  |
| 566 | *Leptotes babaulti* Stempffer, 1935 | GUI | # | + | ?? | bi |  | 3 |  |  |  |  |
| 567 | *Leptotes jeanneli* Stempffer, 1935 | GUI | # | + | ?? | bi |  | 3 |  |  |  |  |
| 569 | *Leptotes pulchra* Murray, 1874 | SPE |  |  | ?? | ?? |  | 4 |  |  |  |  |
| 570 | *Tuxentius cretosus nodieri* Oberthür, 1883 | SUD | # | + | x | x |  | 3 |  |  |  |  |
| 571 | *Tuxentius carana carana* Hewitson, 1876 | ALF | # | + | AN | BI | ı | 2 | DRR |  | KRO | RAP |
| 573 | *Tarucus ungemachi* Stempffer, 1944 | SUD | # | + | x | x |  | 3 |  |  |  |  |
| 579 | *Actizera lucida* Trimen, 1883 | GUI |  | + | x | x |  | 4 |  |  |  |  |
| 580 | *Eicochrysops hippocrates* Fabricius, 1793 | SPE | # | + | AN | BI |  | 2 |  | BOT | KRO | RAP |
| 581 | *Eicochrysops dudgeoni* Riley, 1929 | GUI | # | + | x | x |  | 3 |  |  |  |  |
| 582 | *Cupidopsis jobates jobates* Hopffer, 1855 | SUD |  | + | x | x |  | 3 |  |  |  |  |
| 583 | *Cupidopsis cissus* Godart, 1824 | GUI | # | + | AN | bi |  | 3 |  |  |  |  |
| 585 | *Euchrysops albistriata greenwoodi* D'A., 1980 | GUI | # | + | x | bi |  | 3 |  |  |  |  |
| 587 | *Euchrysops reducta* Hulstaert, 1924[WNE] | GUI | # | + | x | x |  | 4 |  |  |  |  |
| 588 | *Euchrysops malathana* Boisduval, 1833 | UBQ | # | + | AN | BI | ı | 2 |  |  | KRO | RAP |
| 591 | *Euchrysops osiris* Hopffer, 1855 | UBQ | # | + | an | bi |  | 2 |  |  |  |  |
| 592 | *Euchrysops barkeri* Trimen, 1893 | GUI | # | + | x | bi |  | 4 |  |  |  |  |
| 593 | *Euchrysops sabelianus* Libert, 2001[WNE] | SUD | # | + | x | x |  | 4 |  |  |  |  |
| 594 | *Lepidochrysops victoriae victoriae* Ka., 1895 | GUI | # | + | x | x |  | 4 |  |  |  |  |
| 595 | *Lepidochrysops parsimon* Fabricius, 1793 | DRF | # |  | x | bi |  | 4 |  |  |  |  |
| 597 | *Lepidochrysops synchrematiza* B.-B., 1923[WAE] | GUI | # |  | x | bi |  | 4 |  |  |  |  |
| 601 | *Lepidochrysops phasma* Butler, 1901 | GUI | # |  | x | bi |  | 4 |  |  |  |  |
| 602 | *Lepidochrysops quassi* Karsch, 1895[GHE] | GUI | # |  | x | x |  | 3 |  |  |  |  |
| 605 | *Thermoniphas micylus micylus* Cramer, 1780 | MEF | # | + | AN | BI | ı | 2 |  | BOT |  | RAP |
| 609 | *Oboronia punctatus* Dewitz, 1879 | MEF | # | + | AN | BI |  | 2 | DRR |  | KRO | RAP |
| 610 | *Oboronia liberiana* Stempffer, 1950[WAE] | WEF | # | + | AN | BI | ı | 3 | DRR | BOT | KRO | RAP |
| 612 | *Oboronia guessfeldi* Dewitz, 1879 | DRF | # | + | ?? | BI | ı | 3 | DRR |  | KRO | RAP |
| 613 | *Oboronia ornata ornata* Mabille, 1890 | ALF | # | + | AN | BI | ı | 2 | DRR | BOT | KRO | RAP |

A biological assessment of the terrestrial ecosystems of the Draw River, Boi-Tano, Tano Nimiri and Krokosua Hills forest reserves, southwestern Ghana

111

| FAMILY | # | Species name | Habitat choice | collected | | Ankasa / Bia / Krokosua | | | Status | RAP collection localities | | | RAP total |
|---|---|---|---|---|---|---|---|---|---|---|---|---|---|
| | 614 | *Azanus ubaldus* Cramer, 1782 | SUD | | | x | x | | 2 | | | | |
| | 615 | *Azanus jesous* Guérin-Méneville, 1847 | SUD | # | + | x | x | | 2 | | | | |
| | 616 | *Azanus moriqua* Wallengren, 1857 | SUD | # | + | x | bi | | 2 | | | | |
| | 617 | *Azanus mirza* Plötz, 1880 | UBQ | # | + | AN | BI | | 2 | DRR | BOT | KRO | RAP |
| | 618 | *Azanus natalensis* Trimen, 1887 | GUI | | | x | ?? | | 4 | | | | |
| | 619 | *Azanus isis* Drury, 1773 | ALF | # | + | AN | BI | | 2 | | | | |
| | 620 | *Chilades eleusis* Demaison, 1888 | SUD | | | x | x | | 3 | | | | |
| | 621 | *Chilades trochylus* Freyer, 1844 | GUI | # | + | x | bi | | 3 | | | | |
| | 622 | *Zizeeria knysna* Trimen, 1862 | UBQ | # | + | AN | BI | | 2 | | | KRO | RAP |
| | 623 | *Zizina antanossa* Mabille, 1877 | GUI | # | + | AN | BI | | 3 | DRR | | | RAP |
| | 624 | *Zizula hylax* Fabricius, 1775 | UBQ | # | + | AN | BI | | 2 | | | | |
| **RIODINIDAE** | | | | | | | | | | | | | |
| Nemeobiinae | | | | | | | | | | | | | |
| | 625 | *Abisara intermedia* Aurivillius, 1895 | WEF | | | ?? | ?? | | 5 | | | | |
| | 626 | *Abisara tantalus tantalus* Hewitson, 1861 | WEF | | | ?? | ?? | | 5 | | | | |
| | 629 | *Abisara gerontes gerontes* Fabricius, 1781 | WEF | | | ?? | ?? | | 3 | | | | |
| **NYMPHALIDAE** | | | | | | | | | | | | | |
| Libytheinae | | | | | | | | | | | | | |
| | 632 | *Libythea labdaca labdaca* Westwood, 1851 | ALF | # | + | AN | BI | ˋ | 2 | | | KRO | RAP |
| Danainae | | | | | | | | | | | | | |
| | 633 | *Danaus chrysippus chrysippus* Linné, 1758 | UBQ | # | + | AN | BI | ˋ | 1 | | BOT | KRO | RAP |
| | 634 | *Tirumala petiverana* Doubleday & Hew., 1847 | GUI | # | + | an | BI | | 2 | | BOT | | RAP |
| | 636 | *Amauris niavius niavius* Linné, 1758 | GUI | # | + | AN | BI | | 2 | | BOT | | RAP |
| | 637 | *Amauris tartarea tartarea* Mabille, 1876 | ALF | # | + | AN | BI | | 3 | | | | |
| | 638 | *Amauris becate becate* Butler, 1866 | MEF | # | + | an | bi | | 3 | | | | |
| | 639 | *Amauris damocles* Fabricius, 1793[WNE] | DRF | # | + | AN | BI | | 2 | | | | |
| **Satyrinae** | | | | | | | | | | | | | |
| Melanitini | | | | | | | | | | | | | |
| | 643 | *Gnophodes betsimena parmeno* Doubleday, 1849 | ALF | # | + | an | BI | | 2 | | | KRO | RAP |
| | 644 | *Gnophodes chelys* Fabricius, 1793 | MEF | # | + | AN | BI | | 2 | | | KRO | RAP |

Checklist of the Butterflies of Ghana with a list of those collected during the
RAP survey in western Ghana, including those previously known from the
adjacent National Parks of Ankasa and Bia

| FAMILY | No. | Species name | Habitat choice | collected # | collected + | Ankasa | Bia / Krokosua | Krokosua | Status | RAP collection localities | | | RAP total |
|---|---|---|---|---|---|---|---|---|---|---|---|---|---|
| | 645 | Melanitis leda Linné, 1758 | UBQ | # | + | AN | BI | | 2 | | | | |
| | 646 | Melanitis libya Distant, 1882 | UBQ | # | + | AN | bi | | 3 | | | | |
| Elymniini | | | | | | | | | | | | | |
| | 642 | Elymniopsis bammakoo bammakoo Westwood, 1851 | MEF | # | + | AN | BI | | 2 | DRR | | | RAP |
| | 678 | Bicyclus xeneas occidentalis Condamin, 1961 | ALF | # | + | AN | BI | | 3 | | | | |
| | 680 | Bicyclus evadne elionas Hewitson, 1866 | WEF | # | + | AN | BI | | 3 | | | | |
| | 652 | Bicyclus ephorus ephorus Weymer, 1892 | WEF | # | + | an | bi | | 3 | | | | |
| | 655 | Bicyclus italus Hewitson, 1865[VOL] | MEF | # | + | X | X | | 3 | | | | |
| | 656 | Bicyclus zinebi Butler, 1869[WAE] | MEF | # | + | AN | BI | | 3 | DRR | | KRO | RAP |
| | 657 | Bicyclus sangmelinae Condamin, 1963 | WEF | # | + | AN | BI | | 3 | | | | |
| | 658 | Bicyclus sambulos unicolor Condamin, 1971 | WEF | # | + | AN | BI | | 3 | | | | |
| | 659 | Bicyclus mandanes Hewitson, 1873 | DRF | # | + | ?? | BI | | 3 | DRR | | | RAP |
| | 660 | Bicyclus auricrudus Butler, 1868 | MEF | | + | an | bi | | 3 | | | | |
| | 661 | Bicyclus vulgaris Butler, 1868 | ALF | # | + | AN | BI | ˎ | 1 | DRR | | KRO | RAP |
| | 662 | Bicyclus dorothea dorothea Cramer, 1779 | ALF | # | + | AN | BI | ˎ | 1 | DRR | BOT | KRO | RAP |
| | 663 | Bicyclus sandace Hewitson, 1877 | ALF | # | + | AN | BI | ˎ | 1 | DRR | BOT | KRO | RAP |
| | 664 | Bicyclus martius melas Condamin, 1965 | MEF | # | + | AN | BI | ˎ | 2 | DRR | BOT | KRO | RAP |
| | 666 | Bicyclus istaris Plötz, 1880 | WEF | # | + | an | bi | | 3 | | | | |
| | 668 | Bicyclus sylvicolus Condamin, 1965[VOL] | WEF | | + | X | X | | 3 | | | | |
| | 669 | Bicyclus abnormis Dudgeon, 1909[WAE] | WEF | # | + | an | BI | | 3 | DRR | | KRO | RAP |
| | 670 | Bicyclus madetes madetes Hewitson, 1874 | MEF | # | + | AN | BI | | 3 | DRR | | | RAP |
| | 675 | Bicyclus nobilis Aurivillius, 1893 | WEF | # | + | AN | BI | | 4 | | | KRO | RAP |
| | 676 | Bicyclus ignobilis ignobilis Butler, 1870 | ALF | # | + | AN | BI | | 3 | | | | |
| | 677 | Bicyclus maeseni Condamin, 1971[GHE] | ALF | # | + | an | bi | | 3 | | | | |
| | 681 | Bicyclus trilophus jacksoni Condamin, 1961 | WEF | # | + | AN | X | | 3 | | | | |
| | 683 | Bicyclus dekeyseri Condamin, 1958[WAE] | WEF | # | + | an | ?? | | 4 | | | | |
| | 686 | Bicyclus safitza safitza Hewitson, 1851 | GUI | # | + | X | BI | | 3 | | | | |
| | 687 | Bicyclus campa Karsch, 1893 | GUI | # | | X | X | | 3 | | | | |
| | 688 | Bicyclus angulosa angulosa Butler, 1868 | GUI | # | + | X | ?? | | 2 | | | | |
| | 689 | Bicyclus pavonis Butler, 1876 | GUI | | + | X | X | | 2 | | | | |

| FAMILY | | Species name | Habitat choice | collected # | collected + | Ankasa | Bia / Krokosua | Status | RAP collection localities | RAP total |
|---|---|---|---|---|---|---|---|---|---|---|
| | 690 | *Bicyclus milyas* Hewitson, 1864 | GUI | # | | X | X | 2 | | |
| | 691 | *Bicyclus funebris* Guérin-Méneville, 1844 | DRF | # | + | ?? | BI | 2 | | |
| | 692 | *Bicyclus taenias* Hewitson, 1877 | MEF | # | + | AN | BI | 2 | DRR BOT KRO | RAP |
| | 693 | *Bicyclus uniformis* Bethune-Baker, 1908 | WEF | | + | an | ?? | 4 | | |
| | 697 | *Bicyclus procora* Karsch, 1893 | WEF | # | + | AN | bi | 3 | DRR KRO | RAP |
| | 699 | *Hallelesis halyma* Fabricius, 1793[WAE] | MEF | # | + | AN | BI | 3 | DRR BOT | RAP |
| | 700 | *Henotesia elisi* Karsch, 1893[WAE] | DRF | # | + | X | X | 4 | | |
| | 701 | *Heteropsis peitho* Plötz, 1880 | MEF | | + | an | ?? | 4 | KRO | RAP |
| Satyrini | | | | | | | | | | |
| | 702 | *Ypthima asterope asterope* Klug, 1832 | SUD | | + | X | X | 2 | | |
| | 703 | *Ypthima condamini nigeriae* Kielland, 1982 | GUI | # | + | X | ?? | 3 | | |
| | 704 | *Ypthima antennata cornesi* Kielland, 1982 | GUI | # | | X | ?? | 3 | | |
| | 705 | *Ypthima vuattouxi* Kielland, 1982 | ALF | # | + | X | ?? | 4 | | |
| | 706 | *Ypthima doleta* Kirby, 1880 | ALF | # | + | AN | BI | 1 | DRR BOT KRO | RAP |
| | 708 | *Ypthima papillaris papillaris* Butler, 1888 | GUI | # | + | X | bi | 3 | | |
| | 709 | *Ypthima impura impura* Elwes & Edwards, 1893 | GUI | # | + | X | bi | 4 | | |
| | 711 | *Ypthimomorpha ionia* Hewitson, 1865 | SPE | # | + | an | bi | 3 | | |
| Apaturinae | | | | | | | | | | |
| | 712 | *Apaturopsis cleochares cleochares* Hew., 1873 | MEF | # | + | an | BI | 4 | | |
| **Charaxinae** | | | | | | | | | | |
| Charaxini | | | | | | | | | | |
| | 713 | *Charaxes varanes vologeses* Mabille, 1876 | GUI | # | + | ?? | bi | 2 | | |
| | 714 | *Charaxes fulvescens senegala* van S., 1975 | ALF | # | + | AN | bi | 2 | | |
| | 716 | *Charaxes candiope candiope* Godart, 1824 | GUI | | + | X | bi | 3 | | |
| | 717 | *Charaxes protoclea protoclea* Feist., 1850 | ALF | # | + | AN | BI | 2 | DRR BOT KRO | RAP |
| | 718 | *Charaxes boueti* Feisthamel, 1850 | DRF | # | + | ?? | bi | 3 | | |
| | 719 | *Charaxes cynthia cynthia* Butler, 1865 | ALF | # | + | AN | BI | 2 | | |
| | 720 | *Charaxes lucretius lucretius* Cramer, 1775 | ALF | # | + | AN | BI | 2 | KRO | RAP |
| | 721 | *Charaxes lactetinctus lactetinctus* K., 1892 | GUI | # | + | X | X | 4 | | |
| | 722 | *Charaxes epijasius* Reiche, 1850 | GUI | # | + | X | bi | 2 | | |

| FAMILY | | Species name | Habitat choice | collected # | collected + | Ankasa | Bia | Krokosua | Status | RAP DRR | RAP BOT | RAP KRO | RAP total |
|---|---|---|---|---|---|---|---|---|---|---|---|---|---|
| | 724 | *Charaxes castor castor* Cramer, 1775 | DRF | # | + | AN | bi | ' | 3 | | BOT | | RAP |
| | 725 | *Charaxes brutus brutus* Cramer, 1779 | MEF | # | + | AN | BI | ' | 2 | | BOT | | RAP |
| | 726 | *Charaxes pollux pollux* Cramer, 1775 | MEF | | + | ?? | ?? | | 4 | | | | |
| | 728 | *Charaxes eudoxus eudoxus* Drury, 1782 | ALF | | | ?? | ?? | | 4 | | | | |
| | 729 | *Charaxes tiridates tiridates* Cramer, 1777 | ALF | # | + | AN | BI | | 2 | | BOT | KRO | RAP |
| | 730 | *Charaxes bipunctatus bipunctatus* Roth., 1894 | WEF | # | + | an | BI | | 3 | | | | |
| | 731 | *Charaxes numenes numenes* Hewitson, 1859 | ALF | # | + | AN | BI | | 3 | | | | |
| | 732 | *Charaxes smaragdalis butleri* Roth., 1900 | ALF | # | + | AN | bi | | 3 | | | | |
| | 733 | *Charaxes imperialis imperialis* Butler, 1874 | ALF | | | an | bi | | 4 | | | | |
| | 734 | *Charaxes ameliae doumeti* Henning, 1989 | ALF | # | + | AN | BI | | 3 | | | | |
| | 735 | *Charaxes pythodoris occidens* Hewitson, 1873 | DRF | # | | X | bi | | 4 | | | | |
| | 736 | *Charaxes hadrianus hadrianus* Ward, 1871 | WEF | # | + | AN | X | | 4 | | | | |
| | 738 | *Charaxes nobilis claudei* le Moult, 1933 | WEF | | | an | X | | 4 | | | | |
| | 740 | *Charaxes fournierae jolybouyeri* Ving, 1998 | WEF | | | ?? | X | | 4 | | | | |
| | 741 | *Charaxes zingha* Stoll, 1780 | MEF | # | + | AN | bi | | 3 | | BOT | | RAP |
| | 742 | *Charaxes etesipe etesipe* Godart, 1824 | DRF | # | + | AN | bi | | 3 | | | | |
| | 743 | *Charaxes achaemenes atlantica* van S., 1970 | GUI | # | + | X | bi | | 2 | | | | |
| | 744 | *Charaxes eupale* Drury, 1782 | ALF | # | + | AN | BI | ' | 2 | DRR | BOT | KRO | RAP |
| | 745 | *Charaxes subornatus couilloudi* Plant., 1976 | WEF | # | + | an | bi | | 4 | | | | |
| | 746 | *Charaxes anticlea anticlea* Drury, 1782 | ALF | # | + | an | bi | | 3 | | | | |
| | 747 | *Charaxes hildebrandti gillesi* Plant., 1973 | MEF | | + | an | bi | | 4 | | | KRO | RAP |
| | 748 | *Charaxes etheocles etheocles* Cramer, 1777 | ALF | # | + | AN | BI | ' | 2 | DRR | BOT | KRO | RAP |
| | 750 | *Charaxes petersi* van Someren, 1969[WAE] | MEF | | | an | bi | | 4 | | | | |
| | 755 | *Charaxes virilis virilis* van S. & J., 1952 | MEF | # | + | an | bi | | 3 | | | | |
| | 756 | *Charaxes cedreatis* Hewitson, 1874 | MEF | # | | an | bi | | 3 | | | | |
| | 758 | *Charaxes viola* Butler, 1865 | SUD | # | + | X | X | | 2 | | | | |
| | 759 | *Charaxes northcotti* Rothschild, 1899 | GUI | # | | X | bi | | 4 | | | | |
| | 760 | *Charaxes pleione pleione* Godart, 1824 | ALF | # | + | AN | BI | ' | 2 | | | KRO | RAP |
| | 761 | *Charaxes paphianus falcata* Butler, 1872 | WEF | # | + | an | bi | | 3 | | | | |
| | 762 | *Charaxes nichetes bouchei* Plantrou, 1974 | DRF | # | | ?? | bi | | 4 | | | | |
| | 763 | *Charaxes porthos gallayi* van Someren, 1968 | MEF | | | an | bi | | 4 | | | | |

| FAMILY | | Species name | Habitat choice | collected | | Ankasa | Bia / Krokosua | | Status | RAP collection localities | | | RAP total |
|---|---|---|---|---|---|---|---|---|---|---|---|---|---|
| | | | | # | + | | | | | DRR | BOT | KRO | |
| | 764 | *Charaxes zelica zelica* Butler, 1869 | WEF | # | | an | bi | | 4 | | | | |
| | 765 | *Charaxes lycurgus lycurgus* Fabricius, 1793 | ALF | # | + | AN | bi | | 2 | | | KRO | RAP |
| | 766 | *Charaxes mycerina mycerina* Godart, 1824 | WEF | | + | an | ?? | | 4 | | | | |
| | 767 | *Charaxes doubledayi* Aurivillius, 1898 | WEF | | | an | ?? | | 4 | | | | |
| Euxanthini | | | | | | | | | | | | | |
| | 768 | *Euxanthe eurinome eurinome* Cramer, 1775 | ALF | | + | AN | bi | | 3 | DRR | | | RAP |
| Pallini | | | | | | | | | | | | | |
| | 771 | *Palla violinitens* Crowley, 1890 | MEF | # | + | an | BI | | 3 | | | KRO | RAP |
| | 772 | *Palla decius* Cramer, 1777 | MEF | # | + | an | bi | | 3 | | BOT | | RAP |
| | 773 | *Palla ussheri ussheri* Butler, 1870 | ALF | # | + | AN | BI | ʼ | 2 | | | KRO | RAP |
| | 774 | *Palla publius publius* Staudinger, 1892 | MEF | # | + | an | bi | | 3 | | | KRO | RAP |
| **Limenitidinae** | | | | | | | | | | | | | |
| | 777 | *Euryphura togoensis* Suffert, 1904[WNE] | WEF | # | + | an | BI | | 3 | | | | |
| | 779 | *Euryphura chalcis* Felder, 1860 | ALF | # | + | AN | BI | | 2 | DRR | BOT | | RAP |
| | 782 | *Hamanumida daedalus* Fabricius, 1775 | GUI | # | + | AN | BI | | 2 | | BOT | KRO | RAP |
| | 783 | *Aterica galene galene* Brown, 1776 | ALF | # | + | AN | BI | ʼ | 2 | DRR | BOT | KRO | RAP |
| | 784 | *Cynandra opis opis* Drury, 1773 | MEF | # | + | AN | bi | | 3 | | BOT | KRO | RAP |
| | 790 | *Euriphene incerta incerta* Aurivillius, 1912 | MEF | # | + | AN | ?? | | 4 | | | KRO | RAP |
| | 791 | *Euriphene barombina* Aurivillius, 1894 | ALF | # | + | AN | BI | ʼ | 1 | DRR | BOT | | RAP |
| | 792 | *Euriphene veronica* Stoll, 1870[WAE] | WEF | # | + | AN | X | | 2 | DRR | | | RAP |
| | 794 | *Euriphene aridatha transgressa* Hecq, 1994 | MEF | # | + | AN | BI | | 3 | | | | |
| | 795 | *Euriphene ernestibaumanni* Karsch, 1895 | MEF | | | an | ?? | | 4 | | | | |
| | 801 | *Euri. grosesmithi muehlenbergi* Hecq, 1995 | MEF | # | + | an | bi | | 4 | | | | |
| | 806 | *Euriphene coerulea* Boisduval, 1847[WNE] | MEF | # | + | AN | BI | | 2 | DRR | | | RAP |
| | 808 | *Euriphene simplex* Staudinger, 1891[WAE] | WEF | # | + | AN | BI | | 3 | | | | |
| | 815 | *Euriphene larseni* Hecq, 1994[GHE] | MEF | | | X | X | | 5 | | | | |
| | 816 | *Euriphene gambiae vera* Hecq, 2002 | ALF | # | + | AN | BI | | 2 | DRR | | KRO | RAP |
| | 817 | *Euriphene amicia amicia* Hewitson, 1871 | WEF | # | + | an | BI | | 3 | | | KRO | RAP |
| | 818 | *Euriphene* sp. nov. brown Larsen, in prep. | WEF | | | X | X | | 5 | | | | |
| | 819 | *Euriphene ampedusa* Hewitson, 1866[WNE] | ALF | # | + | AN | BI | | 3 | DRR | | KRO | RAP |
| | 820 | *Euriphene leonis* Aurivillius, 1898[WAE] | WEF | | | ?? | X | | 4 | | | KRO | RAP |

| FAMILY | # | Species name | Habitat choice | collected | | Ankasa / Bia / Krokosua | | | Status | RAP collection localities | | | RAP total |
|---|---|---|---|---|---|---|---|---|---|---|---|---|---|
| | | | | # | + | Ankasa | Bia | Krokosua | | DRR | BOT | KRO | |
| | 821 | *Euriphene atossa atossa* Hewitson, 1865 | MEF | # | + | AN | BI | | 3 | | | KRO | RAP |
| | 822 | *Euriphene doriclea doriclea* Drury, 1782 | ALF | # | + | AN | BI | | 3 | | | | RAP |
| | 825 | *Bebearia carshena* Hewitson, 1870 | MEF | # | + | an | bi | | 3 | DRR | | | RAP |
| | 826 | *Bebearia abesa abesa* Hewitson, 1869 | MEF | # | + | an | BI | | 3 | | | KRO | RAP |
| | 827 | *Bebearia banksi* Hecq & Larsen, 1998[GHE] | MEF | # | + | ?? | ?? | | 4 | | | | |
| | 828 | *Bebearia languida* Schultze, 1920 | MEF | # | + | AN | bi | | 4 | | | | |
| | 829 | *Bebearia tentyris* Hewitson, 1866 | MEF | # | + | AN | BI | | 2 | | | KRO | RAP |
| | 830 | *Bebearia osyris* Schultze, 1920[WAE] | MEF | # | + | AN | BI | | 3 | DRR | | | RAP |
| | 832 | *Bebearia absolon absolon* Fabricius, 1793 | ALF | # | + | AN | BI | | 2 | DRR | | KRO | RAP |
| | 834 | *Bebearia zonara* Butler, 1871 | MEF | # | + | ?? | BI | | 2 | | | KRO | RAP |
| | 835 | *Bebearia mandinga mandinga* Felder, 1860 | ALF | # | + | AN | BI | | 2 | | | KRO | RAP |
| | 836 | *Bebearia oxione oxione* Hewitson, 1866 | MEF | # | + | AN | bi | | 3 | | | KRO | RAP |
| | 838 | *Bebearia barce barce* Doubleday, 1847 | WEF | # | + | AN | BI | | 4 | DRR | | | RAP |
| | 839 | *Bebearia sophus phreone* Feisthamel, 1850 | ALF | # | + | AN | BI | | 2 | | BOT | KRO | RAP |
| | 841 | *Bebearia laetitia laetitia* Plötz, 1880 | MEF | # | + | AN | BI | | 2 | | | | |
| | 849 | *Bebearia phantasina* Staudinger, 1891[WNE] | ALF | # | + | AN | BI | | 2 | | | | |
| | 851 | *Bebearia demetra* Godart, 1819 | MEF | # | + | an | BI | | 4 | | | KRO | RAP |
| | 853 | *Bebearia maledicta* Strand, 1912[WNE] | WEF | | + | an | bi | | | | | KRO | RAP |
| | 854 | *Bebearia mardania* Fabricius, 1793[WNE] | ALF | # | + | AN | BI | | 2 | | BOT | KRO | RAP |
| | 855 | *Bebearia cocalia cocalia* Fabricius, 1793 | ALF | # | + | AN | BI | | 2 | DRR | BOT | | RAP |
| | 856 | *Bebearia paludicola paludicola* Holmes, 2000 | MEF | # | + | AN | BI | | 3 | | | KRO | RAP |
| | 861 | *Bebearia arcadius* Fabricius, 1793[WAE] | WEF | # | + | AN | ?? | | 4 | | | | |
| | 865 | *Bebearia cutteri cutteri* Hewitson, 1865 | WEF | # | + | AN | ?? | | 4 | | | | |
| | 869 | *Bebearia ashantina ashantina* Dudgeon, 1913 | WEF | # | + | an | ?? | | 4 | | | | |
| | 876 | *Euphaedra medon medon* Linné, 1763 | ALF | # | + | AN | BI | | 2 | DRR | BOT | KRO | RAP |
| | 877 | *Euphaedra gausape* Butler, 1865[WAE] | WEF | # | + | AN | BI | | 3 | | | KRO | RAP |
| | 885 | *Euphaedra xypete* Hewitson, 1865[WNE] | MEF | # | + | AN | BI | | 2 | DRR | BOT | KRO | RAP |
| | 887 | *Euphaedra hebes* Hecq, 1980[WNE] | WEF | # | + | AN | BI | | 3 | DRR | | | RAP |
| | 889 | *Euphaedra diffusa albocoerulea* Hecq, 1976 | DRF | # | + | X | BI | | 3 | | | | |
| | 890 | *Euphaedra crossei* Sharpe, 1902[WNE] | DRF | # | + | X | X | | 4 | | | | |
| | 891 | *Euphaedra crockeri crockeri* Butler, 1869[WAE] | MEF | # | + | ?? | BI | | 3 | | | KRO | RAP |

| FAMILY | | Species name | Habitat choice | collected | | Ankasa | Bia / Krokosua | Status | RAP collection localities | | | RAP total |
|---|---|---|---|---|---|---|---|---|---|---|---|---|
| | 892 | *Euphaedra eusemoides* S. & K., 1889[WAE] | WEF | # | | ?? | X | 4 | | | | |
| | 894 | *Euphaedra cyparissa cyparissa* Cramer, 1775 | DRF | # | + | ?? | BI | 3 | | | | |
| | 895 | *Euphaedra sarcoptera sarcoptera* Btl., 1871 | MEF | # | + | an | BI | 3 | | | | |
| | 896 | *Euphaedra themis themis* Hübner, 1806 | DRF | # | + | AN | bi | 3 | | | | |
| | 897 | *Euphaedra laboureana eburnensis* Hecq, 1979[WAE] | WEF | # | + | AN | bi | 4 | | | | |
| | 901 | *Euphaedra minuta* Hecq, 1982[WAE] | WEF | # | + | AN | bi | 4 | | | | |
| | 902 | *Euphaedra modesta* Hecq, 1982[WAE] | WEF | # | + | AN | BI | 3 | | | | |
| | 904 | *Euphaedra janetta janetta* Butler, 1871 | ALF | # | + | AN | BI | 2 | DRR | BOT | KRO | RAP |
| | 906 | *Euphaedra vetusta* Butler, 1871[WAE] | WEF | # | + | an | ?? | 4 | | | | |
| | 907 | *Euphaedra aberrans* Staudinger, 1891[WNE] | WEF | # | + | an | X | 4 | | | | |
| | 912 | *Euphaedra ceres ceres* Fabricius, 1775 | ALF | # | + | AN | BI | 2 | DRR | BOT | KRO | RAP |
| | 914 | *Euphaedra phaethusa* Butler, 1865[WAE] | ALF | # | + | AN | BI | 2 | | | KRO | RAP |
| | 915 | *Euphaedra inanum* Butler, 1873[WAE] | MEF | # | + | ?? | BI | 4 | | | KRO | RAP |
| | 925 | *Euphaedra ignota* Hecq, 1996[GHE] | WEF | # | | X | ?? | 5 | | | | |
| | 932 | *Euphaedra velutina* Hecq, 1997 | WEF | # | + | AN | X | 4 | | | | |
| | 935 | *Euphaedra francina* Godart, 1821[WAE] | WEF | # | + | AN | ?? | 3 | DRR | | | RAP |
| | 938 | *Euphaedra eleus eleus* Drury, 1782 | WEF | # | + | AN | BI | 3 | DRR | | | RAP |
| | 941 | *Euphaedra zampa* Westwood, 1850[WAE] | WEF | # | + | AN | bi | 3 | | | | |
| | 943 | *Euphaedra ruspina* Hewitson, 1865[VOL] | WEF | | + | X | X | 4 | | | | |
| | 944 | *Euphaedra eduardsii* van der Hoeven, 1845 | MEF | # | + | AN | BI | 2 | | BOT | KRO | RAP |
| | 945 | *Euphaedra perseis* Drury, 1773[WAE] | WEF | # | + | AN | BI | 3 | | | KRO | RAP |
| | 946 | *Euphaedra harpalyce harpalyce* Cramer, 1777 | ALF | # | + | AN | BI | 2 | DRR | BOT | KRO | RAP |
| | 947 | *Euphaedra eupalus* Fabricius, 1781[WAE] | WEF | # | + | AN | BI | 4 | | | | |
| | 949 | *Catuna crithea crithea* Drury, 1773 | ALF | # | + | AN | BI | 2 | DRR | BOT | KRO | RAP |
| | 950 | *Catuna niji* Fox, 1965 | WEF | # | + | X | X | 4 | | | KRO | RAP |
| | 951 | *Catuna oberthueri* Karsch, 1894 | ALF | # | + | AN | BI | 1 | DRR | BOT | KRO | RAP |
| | 952 | *Catuna angustatum* Felder, 1867 | MEF | # | + | ?? | BI | 2 | | | KRO | RAP |
| | 953 | *Euptera crowleyi crowleyi* Kirby, 1889 | ALF | | + | an | bi | 4 | | | | |
| | 954 | *Euptera elabontas elabontas* Hewitson, 1870 | ALF | # | + | an | bi | 3 | | BOT | KRO | RAP |
| | 955 | *Euptera dorothea* Bethune-Baker, 1904[WAE] | ALF | # | + | AN | bi | 4 | | | KRO | RAP |
| | 956 | *Euptera zowa* Fox, 1965[WNE] | ALF | # | + | an | bi | 3 | | | | |

| FAMILY | | Species name | Habitat choice | collected | | Ankasa | Bia | Krokosua | Status | RAP collection localities | | | RAP total |
|---|---|---|---|---|---|---|---|---|---|---|---|---|---|
| | 957 | Euptera pluto occidentalis Chovet, 1998 | MEF | | | ?? | ?? | | 5 | | | | |
| | 964 | Pseudathyma falcata Jackson, 1969[WNE] | MEF | # | + | an | bi | | 4 | | | | |
| Limenitidini | 965 | Pseudathyma sibyllina Staudinger, 1901[WNE] | MEF | # | + | an | BI | | 4 | | | | |
| | 968 | Harma theobene theobene Doubleday, 1848 | MEF | # | + | AN | BI | ı | 2 | DRR | | KRO | RAP |
| | 974 | Cymothoe egesta egesta Cramer, 1775 | MEF | # | + | AN | BI | ı | 2 | DRR | | KRO | RAP |
| | 975 | Cymothoe fumana fumana Westwood, 1850 | MEF | # | + | AN | BI | ı | 2 | | | KRO | RAP |
| | 978 | Cymothoe lurida lurida Butler, 1871 | WEF | # | + | an | ?? | | 4 | | | | |
| | 983 | Cymothoe aubergeri Plantrou, 1977[GHE] | MEF | # | + | ?? | ?? | | 3 | | | | |
| | 984 | Cymothoe herminia gongoa Fox, 1965 | MEF | # | + | AN | ?? | | 4 | | | | |
| | 985 | Cymothoe weymeri mulatta Belcastro, 1990 | MEF | # | + | an | BI | | 4 | | | | |
| | 988 | Cymothoe caenis Drury, 1773 | ALF | # | + | AN | BI | ı | 2 | | | KRO | RAP |
| | 990 | Cymothoe althea althea Cramer, 1776[WNE] | MEF | # | + | AN | BI | | 3 | | | | |
| | 992 | Cymothoe jodutta jodutta Westwood, 1850 | WEF | # | + | AN | ?? | | 2 | DRR | BOT | KRX | RAP |
| | 996 | Cymothoe coccinata coccinata Hew., 1874 | MEF | # | + | ?? | bi | ı | 3 | | | | |
| | 997 | Cymothoe mabillei Overlaet, 1944[WAE] | MEF | # | + | an | BI | | 2 | | | KRO | RAP |
| | 1002 | Cymothoe 'sangaris' Godart, 1824 | WEF | # | + | an | ?? | | 3 | | | | |
| | 1003 | Pseudoneptis bugandensis ianthe Hem., 1964 | ALF | # | + | AN | BI | ı | 2 | DRR | BOT | KRO | RAP |
| | 1004 | Pseudacraea eurytus Linné, 1758 | ALF | # | + | AN | BI | | 2 | | | KRO | RAP |
| | 1008 | Pseudacraea bois. boisduvalii Dbl., 1845 | DRF | # | + | AN | bi | | 3 | | | | |
| | 1011 | Pseudacraea lucretia lucretia Cr., 1775 | ALF | # | + | AN | BI | ı | 2 | DRR | | KRX | RAP |
| | 1012 | Pseudacraea warburgi Aurivillius, 1892 | MEF | # | + | ?? | BI | | 3 | | | | |
| | 1013 | Pseudacraea hostilia Drury, 1782[WAE] | WEF | # | + | AN | ?? | | 4 | | | | |
| | 1014 | Pseudacraea semire Cramer, 1779 | ALF | # | + | AN | BI | | 2 | DRR | | KRO | RAP |
| | 1016 | Neptis nemetes nemetes Hewitson, 1868 | ALF | # | + | AN | BI | ı | 2 | | BOT | KRX | RAP |
| Neptini | 1017 | Neptis metella metella Dbl. & Hew., 1850 | ALF | # | + | AN | BI | ı | 2 | | BOT | | RAP |
| | 1019 | Neptis serena serena Overlaet, 1955 | DRF | # | + | ?? | BI | | 2 | | | | |
| | 1020 | Neptis kiriakoffi Overlaet, 1955 | DRF | # | + | AN | bi | | 3 | | | | |
| | 1021 | Neptis morosa Overlaet, 1955 | GUI | # | + | ?? | BI | | 2 | | | | |
| | 1022 | Neptis loma Condamin, 1971[WAE] | MEF | | | ?? | ?? | | 4 | | | | |

A biological assessment of the terrestrial ecosystems of the Draw River, Boi-Tano, Tano Nimiri and Krokosua Hills forest reserves, southwestern Ghana

119

| FAMILY | | Species name | Habitat choice | collected | | Ankasa | Bia / Krokosua | | Status | RAP collection localities | | | RAP total |
|---|---|---|---|---|---|---|---|---|---|---|---|---|---|
| | 1023 | *Neptis constantiae angusta* Condamin,1966 | MEF | | | x | x | | 4 | | | | |
| | 1025 | *Neptis alta* Overlaet, 1955 | MEF | # | + | AN | bi | | 3 | | | | |
| | 1026 | *Neptis seeldrayersi* Aurivillius, 1895 | MEF | # | + | an | bi | | 4 | | | | |
| | 1027 | *Neptis ? metanira* Holland, 1892 | MEF | # | + | AN | BI | | 4 | | | | |
| | 1029 | *Neptis najo* Karsch, 1893[WAE] | MEF | # | + | an | BI | | 4 | | | | |
| | 1030 | *Neptis nicomedes* Hewitson, 1874 | MEF | # | + | AN | BI | | 3 | | | | |
| | 1031 | *Neptis quintilla* Mabille, 1890 | MEF | # | | AN | BI | | 4 | | | | |
| | 1032 | *Neptis strigata strigata* Aurivillius, 1894 | MEF | # | + | an | BI | | 4 | | | | |
| | 1033 | *Neptis paula* Staudinger, 1895[WNE] | WEF | # | + | an | bi | | 4 | | | | |
| | 1038 | *Neptis nysiades* Hewitson, 1868 | MEF | # | + | AN | BI | | 3 | DRR | | KRO | RAP |
| | 1040 | *Neptis nicoteles* Hewitson, 1874 | MEF | # | + | AN | BI | | 2 | DRR | | | RAP |
| | 1041 | *Neptis nicobule* Holland, 1892 | MEF | # | + | AN | bi | | 3 | DRR | | | RAP |
| | 1042 | *Neptis mixophyes* Holland, 1892 | WEF | # | + | AN | BI | | 4 | | | | |
| | 1044 | *Neptis nebrodes* Hewitson, 1874 | MEF | # | + | AN | bi | | 3 | | | | |
| | 1045 | *Neptis trigonophora melicertula* Str., 1912 | MEF | # | + | an | bi | | 3 | | | | |
| | 1047 | *Neptis agouale agouale* Pierre-Baltus, 1978 | ALF | # | + | AN | BI | ʿ | 1 | DRR | BOT | KRO | RAP |
| | 1048 | *Neptis melicerta* Drury, 1773 | MEF | # | + | an | BI | | 3 | DRR | | KRO | RAP |
| | 1049 | *Neptis troundi* Pierre-Baltus, 1978 | ALF | # | + | AN | BI | | 2 | DRR | | KRO | RAP |
| **Cyrestinae** | | | | | | | | | | | | | |
| | 1052 | *Cyrestis camillus camillus* Fabricius, 1781 | ALF | # | + | AN | BI | | 2 | DRR | BOT | KRO | RAP |
| | 1053 | *Sallya occidentalium occidentalium* M., 1876 | ALF | # | + | AN | bi | | 3 | | BOT | | |
| | 1054 | *Sallya boisduvali omissa* Rothschild, 1918 | ALF | | | ?? | bi | | 3 | | | | |
| | 1055 | *Sallya umbrina* Karsch, 1892 | DRF | # | + | x | bi | | 3 | | | | |
| | 1057 | *Sallya amulia amulia* Cramer, 1777 | MEF | | | x | x | | 5 | | | | |
| | 1059 | *Byblia anvatara crameri* Aurivillius, 1894 | UBQ | # | + | AN | BI | ʿ | 2 | DRR | | KRO | RAP |
| | 1060 | *Byblia ilithyia* Drury, 1773 | SUD | | | x | x | | 3 | | | | |
| | 1061 | *Mesoxantha ethosea* Drury, 1782 | MEF | # | + | AN | BI | | 3 | | BOT | | RAP |
| | 1062 | *Ariadne enotrea enotrea* Cramer, 1779 | ALF | # | + | AN | BI | ʿ | 2 | DRR | | KRO | RAP |
| | 1063 | *Ariadne albifascia* Joicey & Talbot, 1921 | ALF | # | + | AN | BI | ʿ | 1 | | | KRX | |
| | 1066 | *Neptidopsis ophione ophione* Cramer, 1779 | ALF | # | + | AN | BI | ʿ | 2 | | BOT | KRO | RAP |
| | 1067 | *Eurytela dryope* Cramer, 1775 | DRF | # | + | AN | BI | ʿ | 3 | | | KRO | RAP |

| FAMILY | | Species name | Habitat choice | collected | Ankasa / Bia / Krokosua | | | Status | RAP collection localities | | | RAP total |
|---|---|---|---|---|---|---|---|---|---|---|---|---|
| | | | | | AN | BI | | | DRR | BOT | KRO | |
| | 1069 | Eurytela biarbas biarbas Drury, 1782 | MEF | # | + | AN | BI | ʔ | 2 | | BOT | KRO | RAP |
| **Nymphalinae** | | | | | | | | | | | | |
| Kallimini | | | | | | | | | | | | |
| | 1070 | Hypolimnas misippus Linné, 1767 | UBQ | # | + | AN | BI | ʔ | 2 | DRR | BOT | KRO | RAP |
| | 1071 | Hypolimnas anthedon anthedon Dbl, 1845 | ALF | # | + | AN | BI | ʔ | 2 | DRR | BOT | KRO | RAP |
| | 1072 | Hypolimnas dinarcha dinarcha Hewitson, 1865 | WEF | # | + | AN | bi | ʔ | 3 | DRR | BOT | | RAP |
| | 1075 | Hypolimnas salmacis salmacis Drury, 1773 | ALF | # | + | AN | BI | ʔ | 2 | DRR | BOT | KRO | RAP |
| | 1077 | Salamis cytora Doubleday, 1847[WAE] | MEF | # | + | ?? | bi | ʔ | 3 | | | KRO | RAP |
| | 1079 | Salamis parhassus parhassus Drury, 1782 | ALF | # | + | AN | BI | ʔ | 2 | DRR | BOT | KRO | RAP |
| | 1080 | Salamis anacardii anacardii Linné, 1758 | DRF | # | + | x | bi | ʔ | 3 | | | | |
| | 1081 | Salamis cacta cacta Fabricius, 1793 | ALF | # | + | AN | BI | ʔ | 2 | | BOT | KRO | RAP |
| | 1082 | Junonia orithya madagascariensis Gue., 1865 | SUD | # | + | x | bi | ʔ | 2 | | | | |
| | 1083 | Junonia oenone oenone Linné, 1758 | UBQ | # | + | AN | BI | ʔ | 1 | | | KRO | RAP |
| | 1084 | Junonia hierta cebrene Trimen, 1870 | SUD | # | + | x | bi | ʔ | 2 | | | | |
| | 1085 | Junonia westermanni westermanni West., 1850 | DRF | # | + | ?? | BI | x | 3 | | | KRX | |
| | 1086 | Junonia hadrope Doubleday, 1847 | DRF | # | + | x | x | | 3 | | | | |
| | 1087 | Junonia sophia sophia Fabricius, 1793 | ALF | # | + | AN | BI | ʔ | 2 | DRR | BOT | KRO | RAP |
| | 1088 | Junonia stygia Aurivillius, 1894 | ALF | # | + | AN | BI | ʔ | 2 | | | KRO | RAP |
| | 1090 | Junonia chorimene Guérin-Méneville, 1844 | GUI | # | + | x | bi | ʔ | 2 | | | | |
| | 1091 | Junonia terea terea Drury, 1773 | ALF | # | + | AN | BI | ʔ | 1 | DRR | BOT | KRO | RAP |
| | 1092 | Precis octavia octavia Cramer, 1777 | GUI | # | + | ?? | bi | ʔ | 3 | | | | |
| | 1093 | Precis antilope Feisthamel, 1850 | GUI | # | + | ?? | BI | ʔ | 3 | | | | |
| | 1096 | Precis ceryne ceryne Roth. & Jordan, 1903 | SPE | # | + | ?? | BI | ʔ | 3 | | | | |
| | 1097 | Precis pelarga Fabricius, 1775 | ALF | # | + | AN | BI | ʔ | 3 | | | | |
| | 1098 | Precis sinuata Plötz, 1880 | WEF | # | + | AN | ?? | ʔ | 3 | DRR | | | RAP |
| | 1101 | Catacroptera cloanthe ligata R. & J., 1903 | GUI | # | + | x | bi | ʔ | 3 | | | | |
| | 1102 | Kallimoides rumia Doubleday, 1850 | ALF | # | + | AN | BI | ʔ | 2 | DRR | BOT | KRO | RAP |
| | 1103 | Kamilla cymodoce Cramer, 1777 | MEF | # | + | AN | BI | ʔ | 3 | | | KRO | RAP |
| | 1104 | Vanessula milca Hewitson, 1873 | MEF | | +? | x | ?? | | 3 | | | | |
| Nymphalini | | | | | | | | | | | | |
| | 1105 | Vanessa cardui cardui Linné, 1758 | UBQ | # | + | AN | BI | | 3 | | | | |

| FAMILY | | Species name | Habitat choice | collected | | Ankasa (AN) | Bia (BI) | Krokosua | Status | DRR | BOT | KRO | RAP total |
|---|---|---|---|---|---|---|---|---|---|---|---|---|---|
| | 1106 | *Antanartia delius delius* Drury, 1782 | MEF | # | + | AN | BI | ı | 2 | DRR | BOT | KRO | RAP |
| **Heliconiinae** | | | | | | | | | | | | | |
| Vagrantini | | | | | | | | | | | | | |
| | 1108 | *Lachnoptera anticlia* Hübner, 1819 | MEF | # | + | AN | BI | ı | 2 | DRR | BOT | KRO | RAP |
| | 1109 | *Phalanta phalantha aethiopica* R. & J., 1903 | DRF | # | + | x | bi | | 2 | | | | RAP |
| | 1110 | *Phalanta eurytis eurytis* Dbl. & Hew., 1847 | MEF | # | + | AN | BI | ı | 2 | DRR | | KRO | RAP |
| Acraeini | | | | | | | | | | | | | |
| | 1112 | *Acraea perenna perenna* Doubleday, 1847 | MEF | # | + | ?? | ?? | | 3 | | | | |
| | 1118 | *Acraea circeis* Drury, 1782 | ALF | # | + | AN | BI | ı | 2 | DRR | | KRO | RAP |
| | 1121 | *Acraea translucida derubescens* Elt., 1912 | MEF | # | + | x | x | | 3 | | | | |
| | 1122 | *Acraea peneleos peneleos* Ward, 1871 | ALF | # | + | AN | BI | | 3 | | | KRO | RAP |
| | 1123 | *Acraea parrhasia parrhasia* Fabricius, 1793 | MEF | # | + | AN | BI | ı | 3 | | BOT | KRO | RAP |
| | 1123 | *Acraea orina* Hewitson, 1874 | MEF | # | + | AN | BI | | 4 | | | | |
| | 1124 | *Acraea pharsalus pharsalus* Ward, 1871 | ALF | # | + | AN | BI | | 2 | | | KRO | RAP |
| | 1125 | *Acraea encedon encedon* Linné, 1758 | UBQ | # | + | AN | BI | | 2 | | | | |
| | 1126 | *Acraea encedana* Pierre, 1976 | SPE | # | + | ?? | ?? | | 3 | | | | |
| | 1127 | *Acraea alciope* Hewitson, 1852 | ALF | # | + | AN | BI | ı | 1 | DRR | | KRO | RAP |
| | 1128 | *Acraea aurivillii aurivillii* Stgr., 1896 | ALF | # | + | AN | BI | ı | 3 | | | | |
| | 1129 | *Acraea jodutta jodutta* Fabricius, 1793 | ALF | # | + | AN | BI | ı | 2 | DRR | | KRO | RAP |
| | 1130 | *Acraea lycoa lycoa* Godart, 1819 | ALF | # | + | AN | BI | ı | 2 | | BOT | KRO | RAP |
| | 1131 | *Acraea serena* Fabricius, 1775 | UBQ | # | + | AN | BI | ı | 2 | | | KRO | RAP |
| | 1132 | *Acraea acerata* Hewitson, 1874 | ALF | | + | an | bi | | 3 | | | | |
| | 1133 | *Acraea althoffi pseudepaea* Dudgeon, 1909 | WEF | # | + | ?? | ?? | | 4 | | | | |
| | 1135 | *Acraea bonasia bonasia* Fabricius, 1775 | ALF | # | + | AN | bi | ı | 2 | | | | |
| | 1140 | *Acraea orestia orestia* Hewitson, 1874 | MEF | # | + | AN | bi | ı | 4 | | | KRX | |
| | 1141 | *Acraea polis* Pierre & Bernaud, 1999 | MEF | # | + | AN | bi | | 3 | | | | |
| | 1142 | *Acraea vesperalis* Grose-Smith, 1890 | WEF | | + | ?? | ?? | | 4 | | | | |
| | 1145 | *Acraea kraka kibi* Usher, 1986 | WEF | | | x | x | | 4 | | | | |
| | 1156 | *Acraea rogersi rogersi* Hewitson, 1873 | WEF | # | + | an | bi | | 3 | | | | |
| | 1147 | *Acraea abdera eginopsis* Aur., 1898[WNE] | MEF | # | + | an | bi | | 4 | | | | |
| | 1149 | *Acraea egina egina* Cramer, 1775 | ALF | # | + | AN | BI | ı | 2 | | BOT | KRO | RAP |
| | 1151 | *Acraea pseudegina* Westwood, 1852 | UBQ | # | + | AN | BI | ı | 2 | DRR | | KRX | RAP |

| FAMILY | Species name | Habitat choice | collected | | Ankasa | Bia | Krokosua | Status | RAP collection localities | | | RAP total |
|---|---|---|---|---|---|---|---|---|---|---|---|---|
| 1152 | *Acraea caecilia caecilia* Fabricius, 1781 | SUD | # | + | x | bi | | 2 | | | | |
| 1153 | *Acraea zetes zetes* Linné, 1758 | DRF | # | + | AN | BI | | 3 | | | | |
| 1154 | *Acraea endoscota* le Doux, 1928 | ALF | # | + | AN | BI | | 4 | | | | |
| 1155 | *Acraea leucographa* Ribbe, 1889 | MEF | # | + | an | BI | | 3 | | | | |
| 1157 | *Acraea quirina quirina* Fabricius, 1781 | ALF | # | + | AN | BI | | 2 | DRR | | | RAP |
| 1158 | *Acraea neobule seis* Feisthamel, 1850 | UBQ | # | + | AN | BI | | 2 | | | | |
| 1159 | *Acraea eugenia* Karsch, 1893[VOL] | DRF | # | + | x | x | | 3 | | | | |
| 1160 | *Acraea camaena* Drury, 1773 | DRF | | + | x | ?? | | 4 | | | | |
| 1161 | *Acraea vestalis vestalis* F. & F., 1865 | ALF | # | + | AN | BI | | 3 | DRR | | | RAP |
| 1162 | *Acraea macaria* Fabricius, 1793 | WEF | # | + | AN | ?? | | 3 | | | | |
| 1163 | *Acraea alcinoe alcinoe* F. & F., 1865 | MEF | # | + | AN | BI | | 2 | | | | |
| 1164 | *Acraea consanguinea sartina* Jordan, 1910 | WEF | # | + | AN | ?? | | 4 | | | | |
| 1165 | *Acraea umbra umbra* Drury, 1782 | MEF | # | + | AN | BI | ı | 3 | | | KRX | |
| 1169 | *Acraea epaea epaea* Cramer, 1779 | ALF | # | + | AN | BI | ı | 2 | DRR | BOT | KRX | RAP |
| **HESPERIIDAE** | | | | | | | | | | | | |
| **Coeliadinae** | | | | | | | | | | | | |
| 1172 | *Coeliades chalybe chalybe* Westwood, 1852 | ALF | # | + | AN | BI | | 2 | | | | |
| 1173 | *Coeliades bixana* Evans, 1940 | WEF | # | + | an | ?? | | 3 | | | | |
| 1175 | *Coeliades libeon* Druce, 1875 | ALF | # | + | an | bi | | 3 | | | | |
| 1176 | *Coeliades forestan forestan* Stoll, 1782 | UBQ | # | + | AN | BI | ı | 2 | | | KRX | |
| 1177 | *Coeliades pisistratus* Fabricius, 1793 | ALF | # | + | AN | BI | | 2 | | | | |
| 1178 | *Coeliades hanno* Plötz, 1879 | MEF | # | + | an | BI | | 3 | | | | |
| 1179 | *Pyrrhiades lucagus* Cramer, 1777[WAE] | SPE | # | + | x | x | | 2 | | | | |
| 1180 | *Pyrrhochalcia iphis* Drury, 1773 | ALF | # | + | AN | BI | | 2 | DRR | BOT | KRO | RAP |
| **Pyrginae** | | | | | | | | | | | | |
| 1181 | *Katreus johnstonii* Butler, 1887 | WEF | # | + | AN | ?? | | 4 | | | | |
| 1182 | *Katreus holocausta* Mabille, 1891 | WEF | # | | AN | x | | 5 | | | | |
| 1184 | *Katreus hollandi* Druce, 1909 | WEF | # | + | AN | BI | | 4 | | | | |
| 1185 | *Celaenorrhinus rutilans* Mabille, 1877 | WEF | # | + | an | x | | 4 | | | | |
| 1185a | *Celaenorrhinus* cf *nigrovenata* sp. nov. | WEF | | | an | x | | 5 | | | | |
| 1187 | *Celaenorrhinus leona* Berger, 1975[WAE] | WEF | # | + | an | bi | | 4 | | | | |

A biological assessment of the terrestrial ecosystems of the Draw River, Boi-Tano, Tano Nimiri and Krokosua Hills forest reserves, southwestern Ghana

123

| No. | Species name | Habitat choice | collected | | Ankasa | Bia | Krokosua | Status | RAP collection localities | | | RAP total |
|---|---|---|---|---|---|---|---|---|---|---|---|---|
| | | | # | + | | | | | DRR | BOT | KRO | |
| 1191 | *Celaenorrhinus ankasa* Larsen, in prep.[GHE] | WEF | # | + | AN | x | | 5 | | | | |
| 1192 | *Celaenorrhinus galenus galenus* F, 1793 | WEF | # | + | AN | BI | | 2 | DRR | | KRO | RAP |
| 1193 | *Celaenorrhinus cf. galenus* | WEF | # | + | an | bi | | 3 | | | | |
| 1194 | *Celaenorrhinus meditrina* Hewitson, 1877 | WEF | # | | an | bi | | 4 | | | | |
| 1195 | *Celaenorrhinus ovalis* Evans, 1937 | WEF | # | | an | bi | | 4 | | | | |
| 1198 | *Celaenorrhinus proxima masseni* Berg, 1976 | ALF | # | + | an | BI | | 3 | | | KRO | RAP |
| 1199 | *Celaenorrhinus plagiatus* Berger, 1976 | MEF | # | + | AN | BI | ˎ | 3 | DRR | | KRX | RAP |
| 1200 | *Tagiades flesus* Fabricius, 1781 | ALF | # | + | AN | BI | ˎ | 2 | DRR | BOT | KRO | RAP |
| 1201 | *Eagris denuba denuba* Plötz, 1879 | MEF | # | + | AN | BI | ˎ | 2 | | | KRO | RAP |
| 1203 | *Eagris decastigma* Mabille, 1891 | WEF | # | + | an | ?? | | 4 | | | | |
| 1204 | *Eagris liberti* Larsen & Collins, in prep. | WEF | # | + | an | ?? | | 4 | | | | |
| 1205 | *Eagris subalbida subalbida* Holland, 1894 | WEF | # | + | AN | BI | | 3 | | | | |
| 1206 | *Eagris bereus quaterna* Mabille, 1889 | MEF | # | + | AN | BI | | 3 | | | | |
| 1207 | *Eagris tetrastigma subolivescens* Hol., 1892 | MEF | # | + | an | BI | | 3 | | | | |
| 1208 | *Calleagris lacteus dannatti* Mabille, 1877 | WEF | # | + | AN | BI | | 3 | | | | |
| 1210 | *Procampta rara* Holland, 1892 | MEF | # | + | an | BI | | 3 | | | KRO | RAP |
| 1211 | *Eretis lugens* Rogenhofer, 1891 | GUI | # | + | x | ?? | | 2 | | | | |
| 1212 | *Eretis plistonicus* Plötz, 1879[WNE] | ALF | # | + | AN | bi | | 3 | | | KRO | RAP |
| 1214 | *Eretis melania* Mabille, 1891 | DRF | # | + | ?? | BI | | 3 | | | KRO | RAP |
| 1215 | *Sarangesa laelius* Mabille, 1877 | GUI | # | + | x | bi | | 3 | | | | |
| 1217 | *Sarangesa tertullianus* Fabricius, 1793 | MEF | # | + | AN | BI | | 3 | DRR | | | RAP |
| 1218 | *Sarangesa majorella* Mabille, 1891 | MEF | # | + | an | BI | | 3 | | | | |
| 1219 | *Sarangesa tricerata tricerata* Mab., 1891 | MEF | # | + | ?? | BI | | 3 | | | | |
| 1220 | *Sarangesa thecla thecla* Plötz, 1879 | ALF | # | + | AN | BI | ˎ | 2 | DRR | | KRX | RAP |
| 1221 | *Sarangesa bouvieri* Mabille, 1877 | DRF | # | + | AN | BI | | 2 | DRR | BOT | | RAP |
| 1222 | *Sarangesa brigida brigida* Plötz, 1879 | MEF | # | + | AN | bi | | 3 | | | | |
| 1223 | *Caprona adelica* Karsch, 1892 | GUI | # | + | x | ?? | | 4 | | | | |
| 1224 | *Caprona pillaana* Wallengren, 1857 | SUD | | | x | x | | 4 | | | | |
| 1225 | *Netrobalane canopus* Trimen, 1864 | GUI | | | x | x | | 4 | | | | |
| 1226 | *Abantis bismarcki* Karsch, 1893 | GUI | | | x | ?? | | 4 | | | | |

| FAMILY | | Species name | Habitat choice | collected | Ankasa | Bia / Krokosua | Status | RAP collection localities | RAP total |
|---|---|---|---|---|---|---|---|---|---|
| | 1227 | *Abantis leucogaster leucogaster* M., 1890 | WEF | + | an | bi | 4 | | |
| | 1228 | *Abantis nigeriana* Butler, 1901[WNE] | GUI | + | x | bi | 3 | | |
| | 1229 | *Abantis pseudonigeriana* Usher, 1984[WNE] | SUD | # | x | x | 4 | | |
| | 1231 | *Abantis lucretia lucretia* Druce, 1909 | DRF | # | x | bi | 4 | | |
| | 1232 | *Abantis elegantula elegantula* Mabille, 1890 | WEF | | ?? | ?? | 4 | | |
| | 1233 | *Abantis ja* Druce, 1909 | WEF | | x | x | 5 | | |
| | 1235 | *Spialia spio* Linné, 1767 | SUD | # + | x | bi | 2 | | |
| | 1236 | *Spialia diomus diomus* Hopffer, 1855 | SUD | # + | x | bi | 2 | | |
| | 1237 | *Spialia dromus* Plötz, 1884 | DRF | # + | x | bi | 2 | | |
| | 1238 | *Spialia ploetzi occidentalis* de Jong, 1977 | ALF | # + | an | bi | 3 | KRO | RAP |
| | 1239 | *Gomalia elma elma* Trimen, 1862 | DRF | # + | ?? | BI | 3 | | |
| **Hesperiinae** | 1244 | *Astictopterus anomoeus* Plötz, 1879 | DRF | # + | x | bi | 3 | | |
| | 1245 | *Astictopterus abjecta* Snellen, 1872 | GUI | # + | x | ?? | 3 | | |
| | 1246 | *Prosopalpus debilis* Plötz, 1879 | MEF | # + | AN | bi | 4 | | |
| | 1247 | *Prosopalpus styla* Evans, 1937 | DRF | # + | ?? | bi | 3 | | |
| | 1248 | *Prosopalpus saga* Evans, 1937 | WEF | # + | AN | BI | 4 | BOT | RAP |
| | 1252 | *Gorgyra aretina* Hewitson, 1878 | ALF | # + | AN | bi | 3 | DRR | RAP |
| | 1253 | *Gorgyra heterochrus* Mabille, 1890 | MEF | # + | AN | bi | 3 | | |
| | 1254 | *Gorgyra mocquerysii* Holland, 1896 | ALF | # + | an | bi | 3 | | |
| | 1255 | *Gorgyra aburae* Plötz, 1879[WNE] | WEF | + | an | ?? | 5 | | |
| | 1257 | *Gorgyra bina* Evans, 1937 | MEF | # + | an | bi | 3 | | |
| | 1258 | *Gorgyra sola* Evans, 1937 | MEF | # + | AN | ?? | 4 | | |
| | 1259 | *Gorgyra afikpo* Druce, 1909 | MEF | + | an | bi | 3 | | |
| | 1260 | *Gorgyra diversata* Holland, 1896 | MEF | # + | AN | BI | 3 | | |
| | 1261 | *Gorgyra bule* Miller, 1964[VOL] | MEF | + | x | x | 5 | | |
| | 1262 | *Gorgyra minima* Holland, 1896 | DRF | # + | x | bi | 3 | | |
| | 1263 | *Gorgyra sara* Evans, 1937 | MEF | # + | AN | BI | 3 | | |
| | 1264 | *Gorgyra subfacatus* Mabille, 1889 | MEF | # + | ?? | bi | 3 | | |
| | 1265 | *Gorgyra pali* Evans, 1937 | MEF | # + | AN | BI | 4 | | |
| | 1267 | *Gyrogra subnotata* Holland, 1894 | MEF | # + | AN | bi | 3 | | |

| FAMILY | Species name | Habitat choice | collected | + | AN | BI | Kr | Status | DRR | BOT | KRX | RAP total |
|---|---|---|---|---|---|---|---|---|---|---|---|---|
| 1269 | *Ceratrichia phocion phocion* F., 1781 | MEF | # | + | AN | BI | ' | 2 | | | KRX | RAP |
| 1270 | *Ceratrichia semilutea* Mabille, 1891 | MEF | # | + | ?? | BI | | 4 | | | | |
| 1271 | *Ceratrichia clara clara* Evans, 1937 | WEF | # | + | an | BI | | 3 | | | KRO | RAP |
| 1273 | *Ceratrichia crowleyi* Riley, 1925[WAE] | WEF | # | + | AN | x | | 4 | | | | |
| 1274 | *Ceratrichia nothus* Fabricius, 1787 | WEF | # | + | AN | BI | | 3 | | | KRO | RAP |
| 1275 | *Ceratrichia argyrosticta argyrosticta* 1879 | WEF | # | + | AN | BI | | 3 | | | | |
| 1276 | *Ceratrichia maesseni* Miller, 1971[WAE] | WEF | # | + | AN | BI | | 4 | | | | |
| 1277 | *Teniorhinus watsoni* Holland, 1892 | MEF | # | + | AN | bi | | 3 | | | | |
| 1278 | *Teniorhinus ignita* Mabille, 1877 | MEF | | + | ?? | bi | | 3 | | | | |
| 1279 | *Pardaleodes incerta murcia* Plötz, 1893 | DRF | # | + | ?? | BI | | 3 | | | | |
| 1280 | *Pardaleodes edipus* Stoll, 1781 | ALF | # | + | AN | BI | ~ | 1 | DRR | BOT | KRO | RAP |
| 1281 | *Pardaleodes sator sator* Westwood, 1852 | MEF | # | + | AN | BI | | 3 | | BOT | KRO | RAP |
| 1282 | *Pardaleodes tibullus tibullus* F., 1793 | MEF | # | + | AN | BI | ~ | 3 | | | KRO | RAP |
| 1283 | *Pardaleodes xanthopeplus* Holland, 1892 | WEF | # | | an | ?? | | 5 | | | | |
| 1285 | *Xanthodisca rega* Mabille, 1889[WNE] | MEF | # | + | ?? | bi | | 3 | | | | |
| 1286 | *Xanthodisca astrape* Holland, 1892 | ALF | # | + | ?? | BI | | 3 | | | KRO | RAP |
| 1289 | *Parosmodes morantii axis* Evans, 1937 | GUI | | + | x | x | | 4 | | | | |
| 1290 | *Parosmodes lentiginosa* Holland, 1896 | ALF | | + | ?? | BI | | 4 | | | | |
| 1291 | *Rhabdomantis galatia* Hewitson, 1868 | MEF | # | + | an | BI | | 3 | | | | |
| 1292 | *Rhabdomantis sosia* Mabille, 1891 | MEF | # | + | an | BI | | 3 | | | | |
| 1293 | *Osmodes laronia* Hewitson, 1868 | ALF | # | + | ?? | BI | ~ | 2 | | | KRO | RAP |
| 1294 | *Osmodes omar* Swinhoe, 1916 | MEF | # | + | AN | BI | | 3 | | | | |
| 1295 | *Osmodes lux* Holland, 1892 | WEF | # | + | an | bi | | 3 | | | | |
| 1297 | *Osmodes thora* Plötz, 1884 | ALF | # | + | AN | BI | ~ | 2 | | | KRO | RAP |
| 1298 | *Osmodes distincta* Holland, 1896 | WEF | | + | an | ?? | | 4 | | | | |
| 1299 | *Osmodes adon* Mabille, 1889 | WEF | # | + | an | BI | | 4 | | | | |
| 1301 | *Osmodes adosus* Mabille, 1889 | WEF | # | + | an | BI | | 3 | | | | |
| 1302 | *Osmodes lindseyi* Miller, 1964 | WEF | # | + | an | BI | | 3 | | | | |
| 1303 | *Osmodes costatus* Aurivillius, 1896 | WEF | # | + | an | ?? | | 5 | | | | |
| 1304 | *Osmodes banghaasi* Holland, 1896 | WEF | # | + | AN | x | | 5 | | | | |
| 1305 | *Osphantes ogouena ogouena* Mabille, 1891 | MEF | # | + | ?? | ?? | | 4 | | | | |

| FAMILY | | Species name | Habitat choice | collected | Ankasa | Bia / Krokosua | Status | RAP collection localities | | | RAP total |
|---|---|---|---|---|---|---|---|---|---|---|---|
| | 1306 | *Paracleros placidus* Plötz, 1879 WAE | MEF | # | + | ?? | BI | 3 | | | | |
| | 1307 | *Paracleros biguttulus* Mabille, 1889 | ALF | # | + | AN | BI | 2 | DRR | | KRO | RAP |
| | 1308 | *Paracleros substrigata* Holland, 1890 | MEF | # | + | ?? | BI | 4 | | | | |
| | 1309 | *Paracleros maesseni* Berger, 1978 | MEF | # | + | ?? | ?? | 3 | | | | |
| | 1310 | *Acleros ploetzi* Mabille, 1890 | ALF | # | + | AN | BI | 2 | | | | |
| | 1311 | *Acleros (mackenii) olaus* Plötz, 1884 | ALF | # | + | AN | bi | 2 | | | | |
| | 1312 | *Acleros nigrapex* Strand, 1912 | MEF | # | + | AN | bi | 3 | | | | |
| | 1313 | *Acleros bala* Berger, mss | MEF | # | | x | x | 4 | | | | |
| | 1314 | *Semalea pulvina* Plötz, 1879 | ALF | # | + | AN | BI | 2 | | | | |
| | 1315 | *Semalea sextilis* Plötz, 1886 | WEF | # | + | AN | bi | 3 | | | | |
| | 1316 | *Semalea atrio* Mabille, 1891 | WEF | # | + | an | BI | 4 | | | | |
| | 1318 | *Semalea arela* Mabille, 1891 | DRF | # | + | ?? | BI | 3 | | | | |
| | 1319 | *Hypoleucis ophiusa ophiusa* Hewitson, 1866 | ALF | # | + | AN | BI | 2 | | | | |
| | 1320 | *Hypoleucis tripunctata tripunctata* M., 1891 | MEF | # | + | an | BI | 3 | | | | |
| | 1321 | *Hypoleucis sophia* Evans, 1937 | WEF | # | + | an | BI | 4 | | BOT | | RAP |
| | 1322 | *Meza indusiata* Mabille, 1891 | MEF | # | + | AN | bi | 3 | DRR | | | RAP |
| | 1323 | *Meza meza* Hewitson, 1877 | ALF | # | + | AN | BI | 1 | DRR | | KRO | RAP |
| | 1324 | *Meza mabea* Holland, 1894 | MEF | # | | ?? | ?? | 4 | | | | |
| | 1325 | *Meza leucophaea bassa* Lindsey & Miller, 1965 | MEF | # | + | AN | BI | 3 | | | | |
| | 1326 | *Meza elba* Evans, 1937 | MEF | # | + | an | BI | 4 | | | | |
| | 1327 | *Meza mabillei* Holland, 1894 | WEF | # | + | AN | BI | 4 | | | | |
| | 1328 | *Meza cybeutes volta* Miller, 1971 | ALF | # | + | AN | bi | 3 | | | | |
| | 1330 | *Paronymus xanthias xanthias* Mabille, 1891 | WEF | # | + | AN | ?? | 4 | | | | |
| | 1332 | *Paronymus ligora* Hewitson, 1876 | WEF | # | + | an | BI | 3 | | | | |
| | 1333 | *Paronymus nevea* Druce, 1910 | WEF | # | + | AN | x | 5 | | | | |
| | 1334 | *Andronymus neander neander* Plötz, 1884 | ALF | # | + | an | BI | 3 | | | | |
| | 1336 | *Andronymus caesar caesar* Fabricius, 1793 | ALF | # | + | AN | BI | 2 | DRR | | | RAP |
| | 1337 | *Andronymus bero* Evans, 1937 | MEF | # | + | an | bi | 3 | | | | |
| | 1338 | *Andronymus belles* Evans, 1937 | MEF | # | + | AN | bi | 3 | | | | |
| | 1339 | *Andronymus evander* Mabille, 1890 | MEF | # | + | AN | bi | 3 | | | | |
| | 1342 | *Zophopetes ganda* Evans, 1937 | DRF | # | | x | bi | 4 | | | | |

A biological assessment of the terrestrial ecosystems of the Draw River, Boi-Tano, Tano Nimiri and Krokosua Hills forest reserves, southwestern Ghana

127

| FAMILY | Species name | Habitat choice | collected | | Ankasa | Bia / Krokosua | Status | RAP collection localities | RAP total |
|---|---|---|---|---|---|---|---|---|---|
| 1343 | Zophopetes cerymica Hewitson, 1867 | ALF | # | + | an | bi | 3 | | |
| 1345 | Zophopetes quaternata Mabille, 1876 | DRF | # | | x | x | 4 | | |
| 1346 | Gamia buchholzi Plötz, 1879 | WEF | # | + | AN | bi | 3 | | |
| 1347 | Gamia shelleyi Sharpe, 1890 | WEF | # | + | AN | bi | 3 | | |
| 1348 | Artitropa comus Stoll, 1782 | MEF | | + | an | bi | 3 | | |
| 1349 | Mopala orma Plötz, 1879 | MEF | # | + | an | BI | 3 | | |
| 1350 | Gretna waga Plötz, 1886 | ALF | # | + | AN | BI | 2 | KRO | RAP |
| 1352 | Gretna cylinda Hewitson, 1877 | ALF | # | + | AN | BI | 4 | | |
| 1355 | Gretna balenge zowa Lindsey & Miller, 1965 | MEF | # | + | an | bi | 5 | | |
| 1356 | Pteroteinon laufella Hewitson, 1868 | ALF | # | + | AN | BI | 2 | | |
| 1357 | Pteroteinon iricolor Holland, 1890 | WEF | # | + | AN | BI | 4 | | |
| 1358 | Pteroteinon laterculus Holland, 1890 | WEF | # | + | an | BI | 4 | | |
| 1360 | Pteroteinon caenira Hewitson, 1867 | ALF | # | + | AN | BI | 2 | | |
| 1361 | Pteroteinon ceucaenira Druce, 1910 | WEF | # | + | AN | BI | 4 | | |
| 1362 | Pteroteinon concaenira Belcas. & Lar., 1996 | WEF | # | + | AN | BI | 3 | | |
| 1363 | Pteroteinon pruna Evans, 1937 | MEF | # | + | an | bi | 4 | DRR | RAP |
| 1364 | Leona binoevatus Mabille, 1891 | WEF | # | + | an | bi | 4 | | |
| 1367 | Leona leonora leonora Plötz, 1879 | WEF | | | an | bi | 4 | | |
| 1369 | Leona stoehri Karsch, 1893 | WEF | # | + | an | bi | 4 | | |
| 1370 | Leona meloui Riley, 1925 | WEF | | | an | bi | 4 | | |
| 1373 | Leona luehderi luehderi Plötz, 1879 | WEF | | + | an | bi | 5 | | |
| 1374 | Caenides soritia Hewitson, 1876 | WEF | | + | AN | bi | 4 | | |
| 1375 | Caenides kangvensis Holland, 1896 | MEF | # | + | An | BI | 3 | | |
| 1376 | Caenides xychus Mabille, 1891 | WEF | # | + | an | ?? | 4 | | |
| 1377 | Caenides benga Holland, 1891 | WEF | | | an | ?? | 4 | | |
| 1378 | Caenides otilia Belcastro, 1990 | WEF | # | + | AN | ?? | 4 | | |
| 1379 | Caenides dacenilla Aurivillius, 1925 | WEF | # | + | AN | ?? | 4 | | |
| 1380 | Caenides dacela Hewitson, 1876 | ALF | # | + | AN | BI | 2 | | |
| 1381 | Caenides dacena Hewitson, 1876 | MEF | # | + | AN | bi | 3 | KRO | RAP |
| 1382 | Caenides bidarioides Aurivillius, 1896 | WEF | # | + | an | BI | 4 | | |

| FAMILY | No. | Species name | Habitat choice | collected | | Ankasa | Bia | Status | RAP collection localities | RAP total |
|---|---|---|---|---|---|---|---|---|---|---|
| | 1383 | *Monza alberta* Holland, 1896 | ALF | # | + | AN | BI | 1 | | |
| | 1384 | *Monza cretacea* Snellen, 1872 | ALF | # | + | AN | BI | 2 | | |
| | 1385 | *Melphina noctula* Druce, 1909 | WEF | # | + | AN | BI | 4 | | |
| | 1386 | *Melphina melphis* Holland, 1894 | WEF | | | an | ?? | 4 | | |
| | 1387 | *Melphina unistriga* Holland, 1894 | MEF | # | + | an | BI | 3 | | |
| | 1388 | *Melphina tarace* Mabille, 1891 | MEF | | | an | bi | 4 | | |
| | 1389 | *Melphina flavina* Lindsey & Miller, 1965 | MEF | # | | an | bi | 4 | | |
| | 1390 | *Melphina statirides* Holland, 1896 | MEF | | + | an | bi | 3 | | |
| | 1391 | *Melphina statira* Mabille, 1891 | WEF | | | an | bi | 4 | | |
| | 1393 | *Melphina malthina* Hewitson, 1876 | WEF | # | + | AN | bi | 4 | | |
| | 1394 | *Melphina maximiliani* Belcas. & Lar., prep. | MEF | | + | ?? | ?? | 4 | | |
| | 1395 | *Fresna netopha* Hewitson, 1878 | DRF | # | + | x | ?? | 3 | | |
| | 1396 | *Fresna maesseni* Miller, 1971 [WNE] | MEF | | | x | ?? | 4 | | |
| | 1397 | *Fresna nyassae* Hewitson, 1878 | DRF | | + | x | ?? | 4 | | |
| | 1398 | *Fresna cojo* Karsch, 1893 | ALF | # | + | an | bi | 3 | | |
| | 1399 | *Fresna carlo* Evans, 1937 | MEF | # | + | an | bi | 5 | | |
| | 1400 | *Platylesches galesa* Hewitson, 1877 | ALF | # | + | AN | BI | 3 | | |
| | 1402 | *Platylesches moritili* Wallengren, 1857 | GUI | # | + | x | ?? | 3 | | |
| | 1403 | *Platylesches rossi* Belcastro, 1986 [WAE] | DRF | # | | x | ?? | 5 | | |
| | 1405 | *Platylesches picanini* Holland, 1894 | ALF | # | + | AN | BI | 3 | | |
| | 1407 | *Platylesches chamaeleon chamaeleon* M., 1891 | DRF | | | x | bi | 3 | | |
| | 1411 | *Pelopidas mathias* Fabricius, 1798 | UBQ | # | + | AN | BI | 2 | KRO | RAP |
| | 1412 | *Pelopidas thrax inconspicua* Bertolini, 1850 | UBQ | # | | an | BI | 2 | | |
| | 1413 | *Borbo fallax* Gaede, 1916 | GUI | # | + | x | bi | 3 | | |
| | 1414 | *Borbo fanta fanta* Evans, 1937 | GUI | # | + | AN | bi | 3 | KRO | RAP |
| | 1415 | *Borbo perobscura* Druce, 1912 | GUI | # | + | x | BI | 3 | | |
| | 1416 | *Borbo micans* Holland, 1896 | SPE | # | + | an | BI | 4 | | |
| | 1417 | *Borbo borbonica borbonica* Boisduval, 1833 | GUI | # | + | x | bi | 3 | | |
| | 1418 | *Borbo gemella* Mabille, 1884 | GUI | # | + | x | BI | 3 | | |

A biological assessment of the terrestrial ecosystems of the Draw River, Boi-Tano, Tano Nimiri and Krokosua Hills forest reserves, southwestern Ghana

129

| FAMILY | Species name | Habitat choice | collected | | Ankasa / Bia / Krokosua | | Status | RAP collection localities | RAP total |
|---|---|---|---|---|---|---|---|---|---|
| 1419 | *Borbo binga* Evans, 1937 | MEF | # | + | an | bi | 4 | | |
| 1420 | *Borbo fatuellus fatuellus* Hopffer, 1855 | ALF | # | + | AN | BI | 2 | | |
| 1421 | *Borbo holtzi* Plötz, 1883 | GUI | # | + | x | ?? | 3 | | |
| 1423 | *Parnara monasi* Trimen, 1889 | GUI | # | + | x | bi | 4 | | |
| 1424 | *Gegenes pumilio' gambica* Mabille, 1878 | SUD | # | | x | x | 4 | | |
| 1426 | *Gegenes niso brevicornis* Plötz, 1884 | GUI | # | + | x | bi | 3 | | |
| 1427 | *Gegenes hottentota* Latreille, 1823 | DRF | # | + | x | bi | 3 | | |

| Species centered on | | Ghana | AN | BI |
|---|---|---|---|---|
| *Absolute numbers* | | | | |
| Wet Evergreen Forest | WEF | 244 | 97 | 81 |
| Moist Evergreen Forest | MEF | 266 | 110 | 129 |
| Drier Forest Types | DRF | 70 | 11 | 19 |
| All Forest Types | ALF | 165 | 132 | 132 |
| Guinea Savannah | GUI | 81 | 12 | 12 |
| Sudan Savannah | SUD | 40 | 1 | 1 |
| Special Habitats | SPE | 7 | 3 | 3 |
| Ubiquitous species | UBQ | 30 | 25 | 25 |
| Total | | 903 | 386 | 402 |
| *Percentages* | | | | |
| Wet Evergreen Forest | WEF | 27.0 | | |
| Moist Evergreen Forest | MEF | 29.4 | | |
| Drier Forest Types | DRF | 7.8 | | |
| All Forest Types | ALF | 18.2 | | |
| Guinea Savannah | GUI | 9.0 | | |
| Sudan Savannah | SUD | 4.4 | | |
| Special Habitats | SPE | 0.8 | | |
| Ubiquitous species | UBQ | 3.3 | | |
| Total | | 100.0 | | |

# Appendix 3

## Families of insects collected from the three sites

*Frederick Ansah*

| Taxon | Family | Draw River | Boi-Tano | Krokosua Hills |
|-------|--------|:----------:|:--------:|:--------------:|
| **Coleoptera** | Carabidae | + | - | - |
| | Cantharidae | - | - | + |
| | Cerambycidae | + | + | + |
| | Chrysomelidae | - | - | + |
| | Cincidellidae | + | - | - |
| | Curculionidae | - | - | + |
| | Elateridae | - | + | + |
| | Dysticidae | + | + | + |
| | Gyrinidae | + | + | + |
| | Lycidae | - | + | - |
| | Passalidae | - | - | + |
| | Scarabeidae | + | + | + |
| | Tenebrionidae | + | - | + |
| **Dictyoptera** | Blattidae | + | - | - |
| | Mantidae | - | + | - |
| | Sibyllidae | + | - | - |
| **Diptera** | Asilidae | - | + | - |
| | Calliphoridae | + | - | - |
| | Culicidae | - | + | + |
| | Diopsidae | - | - | + |
| | Drosophilidae | + | + | + |
| | Muscidae | + | + | + |
| | Simuliidae | + | + | + |
| | Tabanidae | + | + | + |
| | Tachinidae | - | + | + |
| | Tephritidae | + | + | - |
| | Tipulidae | - | + | + |

| Taxon | Family | Draw River | Boi-Tano | Krokosua Hills |
|-------|--------|------------|----------|----------------|
| **Dermaptera** | Forficulidae | - | - | + |
| **Heteroptera** | Alydidae | - | + | |
| | Aphophoridae | - | + | - |
| | Cicadidae | + | - | - |
| | Coreidae | - | - | + |
| | Corixidae | + | + | + |
| | Dictyopharidae | - | + | - |
| | Gerridae | + | + | + |
| | Hydrometridae | + | + | + |
| | Membracidae | - | + | - |
| | Pyrrhocoridae | - | + | - |
| | Reduvidae | + | + | + |
| **Hymenoptera** | Apidae | + | + | + |
| | Anthrophoridae | - | - | + |
| | Braconidae | + | + | - |
| | Eumenidae | - | + | - |
| | Formicidae | + | + | + |
| | Ichneumonidae | + | + | + |
| | Megachilidae | - | - | + |
| | Scoliidae | + | + | + |
| | Vespidae | - | + | - |
| **Neuroptera** | Mymeleontidae | + | - | + |
| **Odonata** | Aeshnidae | - | + | - |
| | Coenagriidae | + | - | - |
| | Libellulidae | + | + | + |
| **Orthoptera** | Acrididae | + | + | + |
| | Conocephalidae | + | + | - |
| | Tettigonidae | + | + | + |
| **Isoptera** | Termitidae | - | + | - |

# Appendix 4

## Species of non-butterfly invertebrates identified from the three study sites

*Frederick Ansah*

| Taxon/Family | Species | Draw River | Boi-Tano | Krokosua Hills |
|---|---|:---:|:---:|:---:|
| **Heteroptera** | | | | |
| Alydidae | *Stenocoris* sp. | - | + | + |
| Aphrophoridae | *Poophilus* sp. | - | + | - |
| Coreidae | *Anoplocnemis* sp. | - | - | + |
| | *Cletomorpha lancigera* | - | - | + |
| | *Mygdonia* sp. | - | - | + |
| | *Plectropoda oblongipes* | - | - | + |
| Corixidae | | + | + | + |
| Cicadidae | | + | - | - |
| Dictyopharidae | *Dictyopharina* sp. | - | + | - |
| Gerridae | | + | + | + |
| Hydrometridae | | + | + | + |
| Membracidae | *Leptocentrus* sp. | - | + | - |
| Pyrrhocoridae | *Dysdercus melanoderes* | - | + | - |
| | *Dysdercus* sp. | - | + | - |
| Reduviidae | *Authenta quadrideus funebris* | + | - | - |
| | *Ectrichodia barbicornis* | + | + | - |
| | *Reduvius annulatus* | + | - | - |
| | *Rhinocoris obtusus* | + | + | + |
| | *Rhinocoris loratus* | - | + | + |
| | *Rhinocoris nitidulus* | - | + | + |
| | *Rhinocoris* sp. | + | - | - |
| | *Petalocheirus rubiginosus* | + | - | - |
| **Coleoptera** | | + | - | - |
| Carabidae | *Tithoes spinicornis* | + | + | - |
| | *Glena fasciata* | - | - | + |
| Cerambycidae | *Glena giraffa* | - | - | + |
| | *Glena imparilis* | - | - | + |

A biological assessment of the terrestrial ecosystems of the Draw River, Boi-Tano, Tano Nimiri and Krokosua Hills forest reserves, southwestern Ghana

133

| Taxon/Family | Species | Draw River | Boi-Tano | Krokosua Hills |
|---|---|---|---|---|
| | *Glena quinquelineata* | - | - | + |
| | *Moecha bittneri* | + | - | - |
| | *Stenotomis pulchra* | - | + | - |
| | *Stenotomis* sp. | - | + | - |
| | *Zographus vegalis* | - | - | + |
| | *Silidus* sp. | - | - | + |
| Cantharidae | *Aspidomorpha cincta* | - | - | + |
| Chrysomelidae | *Aspidomorpha quinquefasciatus* | - | - | + |
| | *Catespus beauvoisi* | + | - | - |
| Cincidellidae | *Rhyncophorus phoenicus* | - | - | + |
| Curculionidae | *Dysticus* sp. | + | + | + |
| Dysticidae | *Prosephus* sp. | - | - | + |
| Elateridae | *Gyrinus* sp. | + | + | + |
| Gyrinidae | *Lycus latissimus* | - | + | - |
| Lycidae | *Lycus* sp. | - | + | - |
| | *Erionomus alterego* | - | - | + |
| Passalidae | *Anachalcus cupreus* | + | - | + |
| Scarabeidae | *Cyphonistes rufocasteneus* | + | + | - |
| | *Pachnoda cordata* | - | - | + |
| | *Xylotrupes centaurs* | - | - | + |
| | *Calostega purpuripennis* | + | - | - |
| Tenebrionidae | | | | |
| **Dictyoptera** | | + | - | - |
| Blattidae | *Polyspidota aeuroginosa* | - | + | - |
| Mantidae | *Sphodromantis aurea* | - | + | - |
| | *Sphodromantis lineola* | + | + | + |
| | *Sibylla limbata* | + | - | - |
| Sibyllidae | | | | |
| **Hymenoptera** | *Apis mellifera* | + | + | + |
| Apidae | *Xylocopa torrida* | - | - | + |
| | *Xylocopa* sp. | - | + | - |
| | *Anthophora virida* | - | - | + |
| Anthophoridae | | + | + | - |
| Braconidae | *Synagris rufopicta* | - | + | - |
| Eumenidae | *Camponotus* sp. | + | + | + |
| Formicidae | *Oecophylla longinoda* | + | + | + |
| | *Palthothyreus* sp. | + | + | + |
| | *Polyrachis laboriosa* | - | + | - |

| Taxon/Family | Species | Draw River | Boi-Tano | Krokosua Hills |
|---|---|---|---|---|
| | *Megachile* sp. | - | - | + |
| Megachilidae | | + | + | + |
| Scoliidae | | - | + | - |
| Vespidae | | + | - | + |
| Ichneumonidae | | | | |
| Isoptera | | - | + | - |
| Termitidae | | | | |
| NEUROPTERA | *Macroleon lynceus* | - | - | + |
| Myrmeleontidae | | | | |
| **Odonata** | *Acanthagyna cylidrata* | - | - | + |
| Aeshnidae | *Sapho ciliata* | + | + | + |
| Coenagrionidae | *Acisoma trifidum* | - | + | - |
| Libellulidae | *Crocothemis erythraea* | + | - | - |
| | *Orthetrum braciale* | + | - | - |
| | *Orthetrum* sp. | + | - | - |
| | *Orthetrum icteromelas* | - | + | - |
| | *Palpopleura lucia* | + | - | - |
| | *Philonomon luminans* | + | - | - |
| | *Trithemis arteriosa* | + | - | - |
| **Dermaptera** | *Apachyus depressus* | - | - | + |
| Forficulidae | *Diasperaticus erythrocephalus* | - | - | + |
| **Diptera** | | - | + | - |
| Asilidae | | + | - | - |
| Calliphoridae | *Culex* sp. | + | + | + |
| Culicidae | | - | - | + |
| Diopsidae | *Drosophila* sp. | + | + | + |
| Drosophilidae | *Musca domestica* | + | + | + |
| Muscidae | *Simulium* sp. | + | + | + |
| Simuliidae | *Syrphus* sp. | - | + | - |
| Syrphidae | *Tabanus* sp. | - | - | + |
| Tabanidae | | - | + | + |
| Tachinidae | | + | - | + |
| Tephritidae | | + | + | + |
| Tipulidae | | | | |
| **Orthoptera** | *Acanthacris ruficornis* | - | + | - |
| Acrididae | *Acrida bicolor* | - | + | - |
| | *Badistica ornata* | - | + | - |
| | *Badistica* sp. | - | + | - |
| | *Cantatops modica* | - | + | - |

| Taxon/Family | Species | Draw River | Boi-Tano | Krokosua Hills |
|---|---|---|---|---|
| | *Cantatops spissus* | - | + | - |
| | *Eucoptacra basidens* | + | + | - |
| | *Eucoptacra* sp. | - | + | - |
| | *Euctyopoda bicolor* | - | + | - |
| | *Holopercna gestaeckeri* | + | + | + |
| | *Tylotropidus* sp. | - | + | - |
| | *Zonocerus variegatus* | - | - | + |
| | *Eyprepocnemis* sp. | - | - | + |
| Conocephalidae | *Gymnoproctus* sp. | + | - | - |
| | *Homorocoryphus nitidulus* | - | + | - |
| | *Vossia* sp. | - | + | - |
| Tettigonidae | *Zabalius* sp. | + | - | - |
| **TOTAL** | 93 | 47 | 59 | 54 |

# Appendix 5

## Locality list of amphibian and reptile records from Draw River (DR), Boi-Tano (BT), and Krokosua Hills (KH)

*Raffael Ernst and Alex Cudjoe Agyei*

| Site | Lat (N) | Long (W) | Description |
|------|---------|----------|-------------|
| DR1 | 5 09' 41.95" | 2 23' 31.16" | small rainforest creek with quarzite bottom and rocky outcrops, foot of small hill |
| DR2 | 5 11' 58.35" | 2 23' 33.59" | fast flowing granitic bottom creek with small pools |
| DR3 | 5 10' 05.36" | 2 23' 37.32" | swampy area in vicinity of reserve boundary line, bordered to the left by clearcut area. |
| DR4 | 5 10' 08.53" | 2 23' 37.42" | small partially stagnant quarzite bottom creek, bordered by dense vegetation and extensive swampy area |
| DR5 | 5 11' 47.67" | 2 23' 49.12" | dry closed canopy forest uphill area app. 500 m away from next creek |
| DR6 | 5 11' 47.54" | 2 23' 53.12" | bucket trapline |
| DR7 | no data | no data | |
| DR8 | no data | no data | |
| DR9 | 5 11' 34.87" | 2 24' 16.10" | plantation, mainly cocoa and banana/plantains, several swamps and stagnant creek |
| DR10 | 5 11' 35.49" | 2 24' 26.08" | base camp, open area, clearing near village |
| BT1 | 5 31' 55.58" | 2 37' 07.26" | base camp, open area bordered by cocoa plantations |
| BT2 | 5 31' 38.04" | 2 37' 14.43" | stagnant irrigation canal between access road and cocoa plantation extending in swampy area |
| BT3 | 5 31' 30.59" | 2 37' 22.50" | bucket trapline |
| BT4 | 5 31' 19.79" | 2 37' 23.95" | bucket trapline |
| BT5 | 5 31' 08.94" | 2 37' 25.72" | dry closed forest uphill area |
| BT6 | 5 31' 44.22" | 2 37' 31.32" | small creek at bottom of steep descend near reserve boundary line, flooded pool |
| BT7 | 5 31' 13.80" | 2 37' 32.10" | bucket trapline |
| BT8 | 5 31' 09.00" | 2 37' 40.30" | small gravel bottom rainforest creek in closed canopy forest, fast running segments alternating with larger pools |

| Site | Lat (N) | Long (W) | Description |
|------|---------|----------|-------------|
| BT9 | 5 32' 06.16" | 2 37' 46.97" | swampy area near reserve boundary line, bordered to the right by plantation area |
| BT10 | 5 32' 32.75" | 2 38' 06.95" | dry closed forest uphill area |
| BT11 | 5 30' 27.30" | 2 38' 26.34" | medium pond at bottom of steep descend next to access road, open area, no canopy above pond |
| KH1 | 6 37' 01.23" | 2 50' 29.75" | permanent pond between access road and forest edge |
| KH2 | 6 35' 35.39" | 2 50' 36.39" | large waterfilled holes in buttress root, hilltop (480 m a.s.l.), SE of base camp |
| KH3 | 6 36' 33.08" | 2 50' 48.58" | bucket trap line |
| KH4 | 6 36' 31.73" | 2 50' 53.87" | bucket trap line |
| KH5 | 6 36' 50.09" | 2 50' 58.31" | base camp, open area, clearing near village |
| KH6 | 6 36' 36.28" | 2 51' 01.44" | small creek with large stagnant pools, used for irrigation and water supply by local villagers, vicinity to cocoa plantations |
| KH7 | 6 36' 37.04" | 2 51' 02.35" | stagnant pool within creek, dense vegetation |
| KH8 | 6 36' 36.75" | 2 51' 05.17" | steep hill W of base camp, dry forest |
| KH9 | 6 36' 31.73" | 2 51' 28.15" | bucket trap line |
| KH10 | 6 35' 58.18" | 2 52' 24.94" | SW-extension of reserve, closed canopy forest patch between extensive cocoa plantation and settlement |
| KH11 | 6 35' 97.0" | 2 50' 70.5" | steep hill SE (300 m a.s.l.) of base camp, dry forest, granitic outcrops, thick leaf litter |

Presence, distribution, habitat association and conservation status of the amphibians of the Draw River, Boi-Tano and Krokosua Hills Forest reserves

*Raffael Ernst, Alex Cudjoe Agyei and Mark-Oliver Rödel*

SSA = distributed outside West Africa south of the Sahara; WA = endemic to West Africa, UG = endemic to the Upper Guinea forest zone; F = forest specialist; S = savannah species, FB = farmbush species.

| TAXA | Draw River | Boi - Tano | Krokosua Hills | Restricted to | | | Habitat | | |
|---|---|---|---|---|---|---|---|---|---|
| | | | | SSA | WA | UG | F | S | FB |
| **Amphibia - Anura** | | | | | | | | | |
| **Arthroleptidae** | | | | | | | | | |
| *Arthroleptis* sp. X | X | X | X | | | X | X | | X |
| *Arthroleptis* sp. 2 | X | X | X | | | X | X | | X |
| *Arthroleptis* sp. nov. | | | X | | | | | | ? |
| *Cardioglossa leucomystax* | X | X | | | X | | X | | |
| **Astylosternidae** | | | | | | | | | |
| *Astylosternus* sp. (tadpole) | X | | | | | X | X? | | |
| **Bufonidae** | | | | | | | | | |
| *Bufo regularis* | | | X | X | | | | X | X |
| *Bufo maculatus* | X | X | X | X | | | | X | X |
| **Hemisotidae** | | | | | | | | | |
| *Hemisus marmoratus* | | | X | X | | | | X | |
| **Hyperoliidae** | | | | | | | | | |
| *Acanthixalus sonjae* | | | X | | | X | X | | X |
| *Afrixalus dorsalis* | X | X | X | | X | | | | X |
| *Afrixalus nigeriensis* | | | X | | X | | X | | |
| *Hyperolius concolor* | X | X | X | | X | | | | X |
| *Hyperolius fusciventris (burtoni)* | X | X | | | X | | X | | X |
| *Hyperolius laurenti* | X | X | | | | X | X | | |
| *Hyperolius sylvaticus* | | | X | | | | X | | |
| *Hyperolius viridigulosus* | X | X | | | | X | X | | |
| *Leptopelis hyloides* | X | | X | | X | | X | | X |
| *Leptopelis macrotis* | X | | | | | X | X | | |
| *Leptopelis occidentalis* | X | | | | | X | X | | |
| **Pipidae** | | | | | | | | | |
| *Silurana tropicalis* | | | X | | X | | X | | X |
| **Ranidae** | | | | | | | | | |

Appendix 6

Presence, distribution, habitat association and conservation status of the amphibians of the Draw River, Boi-Tano and Krokosua Hills Forest reserves

| TAXA | Draw River | Boi - Tano | Krokosua Hills | Restricted to | | | Habitat | | |
|---|---|---|---|---|---|---|---|---|---|
| | | | | SSA | WA | UG | F | S | FB |
| *Amnirana albolabris* | | X | X | X | | | X | | X |
| *Amnirana occidentalis* | X | | | | | X | X | | |
| *Aubria occidentalis* | | X | | | | X | X | | |
| *Hoplobatrachus occipitalis* | X | | X | X | | | | X | X |
| *Phrynobatrachus accraensis* | X | X | X | | X | | | X | X |
| *Phrynobatrachus alleni* | X | X | X | | X | | X | | |
| *Phrynobatrachus alticola* | X | X | | | | X | | | X |
| *Phrynobatrachus annulatus* | X | X | | | | X | X | | |
| *Phrynobatrachus calcaratus* | | | X | X | | | | | X |
| *Phrynobatrachus ghanensis* | X | X | | | | X | X | | |
| *Phrynobatrachus gutturosus* | | X | X | | X | | X | | X |
| *Phrynobatrachus liberiensis* | X | X | | | | X | X | | X |
| *Phrynobatrachus plicatus* | X | X | X | | X | | X | | X |
| *Ptychadena aequiplicata* | X | X | X | | X | | X | | |
| *Ptychadena longirostris* | X | X | X | X | | | | | X |
| *Ptychadena mascareniensis* | X | | X | X | | | | | X |
| *Ptychadena oxyrhynchus* | X | X | | X | | | | X | |
| **Rhacophoridae** | | | | | | | | | |
| *Chiromantis rufescens* | X | | | X | | | X | | X |
| **Amphibia - Gymnophiona** | | | | | | | | | |
| **Caecilidae** | | | | | | | | | |
| *Geotrypetes seraphini occidentalis* | X | | | | | X | X | | |
| **Total** | **28** | **22** | **23** | | | | | | |

# Appendix 7

Presence, distribution, habitat association and conservation status of the reptiles of the Draw River, Boi-Tano and Krokosua Hills forest reserves

*Raffael Ernst, Alex Cudjoe Agyei and Mark-Oliver Rödel*

For abbreviations compare to Appendix 6.

| TAXA | Draw River | Boi - Tano | Krokosua Hills | Restricted to | | | Habitat | | | CITES & Red List |
|---|---|---|---|---|---|---|---|---|---|---|
| | | | | SSA | WA | UG | F | S | FB | |
| **Reptilia – Sauria** | | | | | | | | | | |
| **Agamidae** | | | | | | | | | | |
| *Agama agama* | 1 | | 1 | 1 | | | | 1 | 1 | |
| **Chamaeleonidae** | | | | | | | | | | |
| *Chamaeleo gracilis* | | | 1 | | 1? | | | 1 | 1 | App 2 |
| **Gekkonidae** | | | | | | | | | | |
| *Hemidactylus fasciatus* | 1 | | 1 | 1 | | | 1 | | | |
| *Hemidactylus muriceus* | 1 | | | 1 | | | 1 | | | |
| *Lygodactylus conraui* | | 1 | 1 | | | | 1 | | | |
| **Scincidae** | | | | | | | | | | |
| *Cophoscincopus durus* | 1 | | | | | 1 | 1 | | | |
| *Mabuya affinis* | 1 | 1 | 1 | | 1 | | 1 | | 1 | |
| *Mabuya polytropis paucisquamis* | 1 | 1 | | | | 1 | 1 | | | |
| *Panaspis togoensis* | 1 | | 1 | | | | | | | |
| *Mochlus fernandi* | 1 | | | | | | | | | |
| **Reptilia – Serpentes** | | | | | | | | | | |
| **Atractaspidae** | | | | | | | | | | |
| *Atractaspis aterrima* | | 1 | | | | | | | | |
| **Colubridae** | | | | | | | | | | |
| *Lycophidion nigromaculatum* | 1 | | | | | 1 | 1 | | | |
| *Toxicodryas pulverulenta* | 1 | | | | 1 | | 1 | | | |
| *Thrasops* sp. | 1 | | | | | | | | | |
| **Elapidae** | | | | | | | | | | |
| *Dendroaspis viridis* | | | 1 | | | 1 | 1 | | | |
| **Viperidae** | | | | | | | | | | |
| *Causus maculatus* | | | 1 | 1 | | | 1 | 1 | 1 | |
| **Reptilia – Crocodylia** | | | | | | | | | | |
| **Crocodylidae** | | | | | | | | | | |
| *Osteolaemus tetraspis* | 1 | 1 | | 1 | | | 1 | | | App 1, VU |
| **Reptilia – Chelonia** | | | | | | | | | | |
| **Testudinidae** | | | | | | | | | | |
| *Kinixys erosa* | 1 | | | 1 | | | 1 | | | DD |
| **Total** | **13** | **5** | **8** | | | | | | | |

# Appendix 8

## Checklist of the birds of Draw River, Boi-Tano and Krokosua Hills forest reserves

*Hugo Rainey and Augustus Asamoah*

| Species | | Draw River | | Boi-Tano | | Krokosua | | Threat Status | Rest. Range | Biome |
|---|---|---|---|---|---|---|---|---|---|---|
| | | Abundance | Breeding | Abundance | Breeding | Abundance | Breeding | | | |
| **ARDEIDAE (1)** | | | | | | | | | | |
| *Tigriornis leucolophus* | White-crested Tiger Heron | | | x | | | | DD | | GC |
| **THRESKIORNITHIDAE (1)** | | | | | | | | | | |
| *Bostrychia olivacea* | Olive Ibis | | | x | | | | | | |
| **ACCIPITRIDAE (12)** | | | | | | | | | | |
| *Pernis apivorus* | European Honey Buzzard | | | R | | | | | | |
| *Gypohierax angolensis* | Palm-nut Vulture | R | | x | | | | | | |
| *Circaetus cinereus* | Brown Snake Eagle | R | | | | | | | | |
| *Dryotriorchis spectabilis* | Congo Serpent Eagle | R | | | | | | | | GC |
| *Polyboroides typus* | African Harrier Hawk | F | | C | | U | | | | |
| *Accipiter tachiro* | African Goshawk | U | | R | | C | | | | |
| *Accipiter erythropus* | Red-thighed Sparrowhawk | R | | x | | | | | | GC |
| *Accipiter melanoleucus* | Black Sparrowhawk | R | | R | | | | | | |
| *Urotriorchis macrourus* | Long-tailed Hawk | x | | x | | x | | | | GC |
| *Buteo auguralis* | Red-necked Buzzard | | | | | R | | | | |
| *Spizaetus africanus* | Cassin's Hawk Eagle | | | R | | R | | | | GC |
| *Stephanoaetus coronatus* | Crowned Eagle | | | R | | U | | | | |
| **PHASIANIDAE (2)** | | | | | | | | | | |
| *Francolinus lathami* | Latham's Forest Francolin | R | | x | | F | | | | GC |
| *Francolinus ahantensis* | Ahanta Francolin | | | R | | F | | | | GC |
| **NUMIDIDAE (1)** | | | | | | | | | | |
| *Agelastes meleagrides* | White-breasted Guineafowl | | | x | | | | VU | RR | GC |
| **RALLIDAE (2)** | | | | | | | | | | |
| *Himantornis haematopus* | Nkulengu Rail | U | | U | | | | | | GC |
| *Sarothrura pulchra* | White-spotted Flufftail | C | | F | | F | | | | GC |
| **SCOLOPACIDAE (1)** | | | | | | | | | | |
| *Actitis hypoleucos* | Common Sandpiper | | | x | | | | | | |

| Species | | Draw River | | Boi-Tano | | Krokosua | | | | |
|---|---|---|---|---|---|---|---|---|---|---|
| | | Abundance | Breeding | Abundance | Breeding | Abundance | Breeding | Threat Status | Rest. Range | Biome |
| **COLUMBIDAE (6)** | | | | | | | | | | |
| *Treron calva* | African Green Pigeon | C | | C | | C | B | | | |
| *Turtur brehmeri* | Blue-headed Wood Dove | C | | C | | C | | | | GC |
| *Turtur tympanistria* | Tambourine Dove | C | | F | | C | | | | |
| *Columba iriditorques* | Western Bronze-naped Pigeon | C | | C | | C | | | | GC |
| *Columba unicincta* | Afep Pigeon | | | x | | U | | | | GC |
| *Streptopelia semitorquata* | Red-eyed Dove | | | | | F | | | | |
| **PSITTACIDAE (2)** | | | | | | | | | | |
| *Psittacus erithacus* | Grey Parrot | R | | x | | U | | | | GC |
| *Poicephalus gulielmi* | Red-fronted Parrot | x | | F | | C | | | | |
| **MUSOPHAGIDAE (3)** | | | | | | | | | | |
| *Corythaeola cristata* | Great Blue Turaco | R | | x | | | | | | |
| *Tauraco persa* | Green Turaco | | | | | x | | | | GC |
| *Tauraco macrorhynchus* | Yellow-billed Turaco | C | | C | | C | | | | GC |
| **CUCULIDAE (11)** | | | | | | | | | | |
| *Cuculus solitarius* | Red-chested Cuckoo | R | | x | | R | | | | |
| *Cuculus clamosus* | Black Cuckoo | F | | F | | R | | | | |
| *Cercococcyx mechowi* | Dusky Long-tailed Cuckoo | R | | U | | | | | | GC |
| *Cercococcyx olivinus* | Olive Long-tailed Cuckoo | C | | C | | R | | | | GC |
| *Chrysococcyx cupreus* | African Emerald Cuckoo | F | | F | | F | | | | |
| *Chrysococcyx flavigularis* | Yellow-throated Cuckoo | | | R | | | | | | GC |
| *Chrysococcyx klaas* | Klaas's Cuckoo | R | | x | | C | | | | |
| *Chrysococcyx caprius* | Didric Cuckoo | | | | | F | | | | |
| *Ceuthmochares aereus* | Yellowbill | C | | C | | C | | | | |
| *Centropus leucogaster* | Black-throated Coucal | C | | C | | F | | | | GC |
| *Centropus grillii* | Black Coucal | | | | | x | | | | |
| **STRIGIDAE (5)** | | | | | | | | | | |
| *Otus icterorhynchus* | Sandy Scops Owl | R | | R | | | | | | GC |
| *Bubo poensis* | Fraser's Eagle Owl | R | | | | R | | | | GC |
| *Bubo leucostictus* | Akun Eagle Owl | | | | | R | | | | GC |
| *Glaucidium tephronotum* | Red-chested Owlet | | | | | U | | | | GC |
| *Strix woodfordii* | African Wood Owl | U | | R | | F | | | | |
| **APODIDAE (6)** | | | | | | | | | | |
| *Rhaphidura sabini* | Sabine's Spinetail | F | | R | | U | | | | GC |
| *Telacanthura melanopygia* | Black Spinetail | | | | | R | | | | GC |
| *Neafrapus cassini* | Cassin's Spinetail | R | | | | R | | | | GC |
| *Cypsiurus parvus* | African Palm Swift | | | | | U | | | | |
| *Apus apus* | European Swift | F | | C | | C | | | | |

| Species | | Draw River | | Boi-Tano | | Krokosua | | Threat Status | Rest. Range | Biome |
|---|---|---|---|---|---|---|---|---|---|---|
| | | Abundance | Breeding | Abundance | Breeding | Abundance | Breeding | | | |
| *Apus caffer* | White-rumped Swift | R | | | | | | | | |
| **TROGONIDAE (1)** | | | | | | | | | | |
| *Apaloderma narina* | Narina's Trogon | R | | U | | R | | | | |
| **ALCEDINIDAE (7)** | | | | | | | | | | |
| *Halcyon badia* | Chocolate-backed Kingfisher | C | | C | | R | | | | GC |
| *Halcyon malimbica* | Blue-breasted Kingfisher | U | | F | | | | | | |
| *Halcyon senegalensis* | Woodland Kingfisher | | | x | | C | | | | |
| *Ceyx lecontei* | African Dwarf Kingfisher | x | | x | | | | | | GC |
| *Alcedo leucogaster* | White-bellied Kingfisher | U | | U | | | | | | GC |
| *Alcedo quadribrachys* | Shining-blue Kingfisher | | | x | | | | | | |
| *Megaceryle maxima* | Giant Kingfisher | | | x | | | | | | |
| **MEROPIDAE (3)** | | | | | | | | | | |
| *Merops muelleri* | Blue-headed Bee-eater | | | x | | | | | | GC |
| *Merops gularis* | Black Bee-eater | U | | U | | R | | | | GC |
| *Merops albicollis* | White-throated Bee-eater | U | | U | | F | | | | |
| **CORACIIDAE (1)** | | | | | | | | | | |
| *Eurystomus gularis* | Blue-throated Roller | R | | | | F | | | | GC |
| **PHOENICULIDAE (1)** | | | | | | | | | | |
| *Phoeniculus castaneiceps* | Forest Wood-hoopoe | U | | | | F | | | | GC |
| **BUCEROTIDAE (9)** | | | | | | | | | | |
| *Tropicranus albocristatus* | White-crested Hornbill | F | | U | | F | | | | GC |
| *Tockus hartlaubi* | Black Dwarf Hornbill | R | | x | | | | | | GC |
| *Tockus camurus* | Red-billed Dwarf Hornbill | F | | x | | | | | | GC |
| *Tockus fasciatus* | African Pied Hornbill | C | b | C | | C | | | | GC |
| *Bycanistes fistulator* | Piping Hornbill | C | | C | | | | | | GC |
| *Bycanistes subcylindricus* | Black-and-white-casqued Hornbill | | | x | | | | | | GC |
| *Bycanistes cylindricus* | Brown-cheeked Hornbill | x | | | | R | | NT | RR | GC |
| *Ceratogymna atrata* | Black-casqued Hornbill | x | | | | | | | | GC |
| *Ceratogymna elata* | Yellow-casqued Hornbill | x | | | | | | NT | | GC |
| **CAPITONIDAE (9)** | | | | | | | | | | |
| *Gymnobucco peli* | Bristle-nosed Barbet | R | | C | B | U | | | | GC |
| *Gymnobucco calvus* | Naked-faced Barbet | F | | R | | C | | | | GC |
| *Pogoniulus scolopaceus* | Speckled Tinkerbird | C | | C | | C | | | | GC |
| *Pogoniulus atroflavus* | Red-rumped Tinkerbird | C | | F | | R | | | | GC |
| *Pogoniulus subsulphureus* | Yellow-throated Tinkerbird | C | | F | | C | | | | GC |
| *Pogoniulus bilineatus* | Yellow-rumped Tinkerbird | F | | | | U | | | | |
| *Buccanodon duchaillui* | Yellow-spotted Barbet | C | | C | | C | | | | GC |

| Species | | Draw River | | Boi-Tano | | Krokosua | | Threat Status | Rest. Range | Biome |
|---|---|---|---|---|---|---|---|---|---|---|
| | | Abundance | Breeding | Abundance | Breeding | Abundance | Breeding | | | |
| *Tricholaema hirsuta* | Hairy-breasted Barbet | C | | C | | C | | | | GC |
| *Trachylaemus purpuratus* | Yellow-billed Barbet | C | | C | | C | | | | GC |
| **INDICATORIDAE (4)** | | | | | | | | | | |
| *Prodotiscus insignis* | Cassin's Honeybird | R | | | | | | | | GC |
| *Melichneutes robustus* | Lyre-tailed Honeyguide | | | x | | | | | | GC |
| *Indicator maculatus* | Spotted Honeyguide | x | | R | | | | | | GC |
| *Indicator willcocksi* | Willcocks's Honeyguide | x | | | | | | | | GC |
| **PICIDAE (5)** | | | | | | | | | | |
| *Campethera maculosa* | Little Green Woodpecker | | | x | | | | | | GC |
| *Campethera nivosa* | Buff-spotted Woodpecker | F | | U | | R | | | | GC |
| *Campethera caroli* | Brown-eared Woodpecker | R | | R | | | | | | GC |
| *Dendropicos gabonensis* | Gabon Woodpecker | R | | R | | | | | | GC |
| *Dendropicos pyrrhogaster* | Fire-bellied Woodpecker | F | | x | | C | | | | GC |
| **EURYLAIMIDAE (1)** | | | | | | | | | | |
| *Smithornis rufolateralis* | Rufous-sided Broadbill | C | b | C | | C | | | | GC |
| **HIRUNDINIDAE (6)** | | | | | | | | | | |
| *Psalidoprocne nitens* | Square-tailed Saw-wing | R | | U | | F | | | | GC |
| *Psalidoprocne obscura* | Fanti Saw-wing | | | x | | | | | | GC |
| *Hirundo semirufa* | Rufous-chested Swallow | | | | | F | | | | |
| *Hirundo abyssinica* | Lesser Striped Swallow | | | | | F | | | | |
| *Hirundo nigrita* | White-throated Blue Swallow | x | | x | | | | | | GC |
| *Hirundo rustica* | Barn Swallow | F | | R | | C | | | | |
| **CAMPEPHAGIDAE (1)** | | | | | | | | | | |
| *Coracina azurea* | Blue Cuckoo-shrike | R | | C | | C | | | | GC |
| **PYCNONOTIDAE (21)** | | | | | | | | | | |
| *Andropadus virens* | Little Greenbul | C | | C | | C | | | | |
| *Andropadus gracilis* | Little Grey Greenbul | U | | x | | R | | | | GC |
| *Andropadus ansorgei* | Ansorge's Greenbul | | | | | R | | | | GC |
| *Andropadus curvirostris* | Cameroon Sombre Greenbul | F | | F | | U | | | | GC |
| *Andropadus gracilirostris* | Slender-billed Greenbul | C | | C | | C | | | | |
| *Andropadus latirostris* | Yellow-whiskered Greenbul | C | | C | | C | | | | |
| *Calyptocichla serina* | Golden Greenbul | x | | | | R | | | | GC |
| *Baeopogon indicator* | Honeyguide Greenbul | C | | C | | C | | | | GC |
| *Ixonotus guttatus* | Spotted Greenbul | U | | U | | | | | | GC |
| *Chlorocichla simplex* | Simple Leaflove | R | | R | | F | | | | GC |
| *Thescelocichla leucopleura* | Swamp Palm Bulbul | C | | | | R | | | | GC |
| *Phyllastrephus icterinus* | Icterine Greenbul | C | | C | | C | | | | GC |

| Species | | Draw River | | Boi-Tano | | Krokosua | | Threat Status | Rest. Range | Biome |
|---------|--|:---------:|:--------:|:---------:|:--------:|:---------:|:--------:|:-------------:|:-----------:|:-----:|
| | | Abundance | Breeding | Abundance | Breeding | Abundance | Breeding | | | |
| *Phyllastrephus albigularis* | White-throated Greenbul | | | | | U | | | | GC |
| *Bleda syndactyla* | Red-tailed Bristlebill | C | | C | | C | | | | GC |
| *Bleda eximia* | Green-tailed Bristlebill | U | | U | | R | | VU | RR | GC |
| *Bleda canicapilla* | Grey-headed Bristlebill | C | | F | | C | | | | GC |
| *Criniger barbatus* | Western Bearded Greenbul | C | | C | | C | | | | GC |
| *Criniger calurus* | Red-tailed Greenbul | C | | C | | C | | | | GC |
| *Criniger olivaceus* | Yellow-bearded Greenbul | R | | x | | | | VU | RR | GC |
| *Pycnonotus barbatus* | Common Bulbul | | | x | | C | | | | |
| *Nicator chloris* | Western Nicator | C | | C | | C | | | | GC |
| **TURDIDAE (4)** | | | | | | | | | | |
| *Stiphrornis erythrothorax* | Forest Robin | C | | C | | C | | | | GC |
| *Alethe diademata* | Fire-crested Alethe | C | | R | | C | | | | GC |
| *Neocossyphus poensis* | White-tailed Ant Thrush | F | | U | | x | | | | GC |
| *Stizorhina finschi* | Finsch's Flycatcher Thrush | C | | C | | C | | | | GC |
| **SYLVIIDAE (14)** | | | | | | | | | | |
| *Cisticola lateralis* | Whistling Cisticola | | | | | U | | | | |
| *Prinia subflava* | Tawny-flanked Prinia | | | | | F | | | | |
| *Apalis nigriceps* | Black-capped Apalis | F | | U | | R | | | | GC |
| *Apalis sharpii* | Sharpe's Apalis | C | | C | | C | | | RR | GC |
| *Camaroptera brachyura* | Grey-backed Camaroptera | | | | | C | | | | |
| *Camaroptera superciliaris* | Yellow-browed Camaroptera | F | | C | | C | | | | GC |
| *Camaroptera chloronota* | Olive-green Camaroptera | F | | F | | C | | | | GC |
| *Macrosphenus kempi* | Kemp's Longbill | x | | | | F | | | | GC |
| *Macrosphenus concolor* | Grey Longbill | C | | C | | C | | | | GC |
| *Eremomela badiceps* | Rufous-crowned Erememela | | | R | | U | | | | GC |
| *Sylvietta virens* | Green Crombec | R | | x | | | | | | GC |
| *Sylvietta denti* | Lemon-bellied Crombec | | | | | U | | | | GC |
| *Hyliota violacea* | Violet-backed Hyliota | | | x | | R | | | | GC |
| *Hylia prasina* | Green Hylia | C | | C | | C | | | | GC |
| **MUSCICAPIDAE (8)** | | | | | | | | | | |
| *Fraseria ocreata* | Fraser's Forest Flycatcher | U | b | F | | R | | | | GC |
| *Fraseria cinerascens* | White-browed Forest Flycatcher | | | x | | | | | | GC |
| *Muscicapa caerulescens* | Ashy Flycatcher | | | | | R | | | | |
| *Muscicapa cassini* | Cassin's Flycatcher | | | x | | | | | | GC |
| *Muscicapa epulata* | Little Grey Flycatcher | x | | R | | F | | | | GC |
| *Muscicapa comitata* | Dusky-blue Flycatcher | U | | R | | R | | | | GC |
| *Muscicapa ussheri* | Ussher's Flycatcher | U | | F | | U | | | | GC |

| Species | | Draw River | | Boi-Tano | | Krokosua | | | | |
|---|---|---|---|---|---|---|---|---|---|---|
| | | Abundance | Breeding | Abundance | Breeding | Abundance | Breeding | Threat Status | Rest. Range | Biome |
| *Myioparus griseigularis* | Grey-throated Flycatcher | R | | | | R | | | | GC |
| **MONARCHIDAE (4)** | | | | | | | | | | |
| *Erythrocercus mccallii* | Chestnut-capped Flycatcher | R | | | | U | | | | GC |
| *Elminia nigromitrata* | Dusky Crested Flycatcher | x | | | | | | | | GC |
| *Trochocercus nitens* | Blue-headed Crested Flycatcher | U | | R | | R | | | | GC |
| *Terpsiphone rufiventer* | Red-bellied Paradise Flycatcher | C | | C | | C | | | | GC |
| **PLATYSTEIRIDAE (3)** | | | | | | | | | | |
| *Megabyas flammulatus* | Shrike Flycatcher | | | U | | R | | | | GC |
| *Dyaphorophyia castanea* | Chestnut Wattle-eye | C | | C | | C | | | | GC |
| *Dyaphorophyia blissetti* | Red-cheeked Wattle-eye | | | | | x | | | | GC |
| **TIMALIIDAE (4)** | | | | | | | | | | |
| *Illadopsis rufipennis* | Pale-breasted Illadopsis | C | | F | | C | | | | |
| *Illadopsis fulvescens* | Brown Illadopsis | | | | | U | | | | GC |
| *Illadopsis cleaveri* | Blackcap Illadopsis | R | | U | | x | | | | GC |
| *Illadopsis rufescens* | Rufous-winged Illadopsis | x | | U | | R | | NT | RR | GC |
| **REMIZIDAE (1)** | | | | | | | | | | |
| *Pholidornis rushiae* | Tit-hylia | | | R | | | | | | GC |
| **NECTARINIIDAE (10)** | | | | | | | | | | |
| *Anthreptes rectirostris* | Green Sunbird | | | x | | | | | | GC |
| *Deleornis fraseri* | Fraser's Sunbird | U | | C | | U | | | | GC |
| *Cyanomitra cyanolaema* | Blue-throated Brown Sunbird | F | | U | | U | | | | GC |
| *Cyanomitra obscura* | Western Olive Sunbird | C | | C | | C | | | | |
| *Chalcomitra adelberti* | Buff-throated Sunbird | C | | U | | U | | | | GC |
| *Hedydipna collaris* | Collared Sunbird | F | | F | | C | B | | | |
| *Cinnyris chloropygius* | Olive-bellied Sunbird | | | R | | F | | | | |
| *Cinnyris johannae* | Johanna's Sunbird | R | b | U | | | | | | GC |
| *Cinnyris superbus* | Superb Sunbird | R | | x | | C | | | | GC |
| *Cinnyris batesi* | Bates's Sunbird | R | | R | | | | | | GC |
| **MALACONOTIDAE (1)** | | | | | | | | | | |
| *Dryoscopus sabini* | Sabine's Puffback | x | | R | | C | | | | GC |
| **PRIONOPIDAE (1)** | | | | | | | | | | |
| *Prionops caniceps* | Red-billed Helmet-shrike | F | B | C | | F | | | | GC |
| **ORIOLIDAE (2)** | | | | | | | | | | |
| *Oriolus nigripennis* | Black-winged Oriole | R | | x | | U | | | | GC |
| *Oriolus brachyrhynchus* | Western Black-headed Oriole | C | | C | | C | | | | GC |

| Species | | Draw River | | Boi-Tano | | Krokosua | | Threat Status | Rest. Range | Biome |
|---|---|---|---|---|---|---|---|---|---|---|
| | | Abundance | Breeding | Abundance | Breeding | Abundance | Breeding | | | |
| **DICRURIDAE (2)** | | | | | | | | | | |
| *Dicrurus atripennis* | Shining Drongo | U | | C | B | C | | | | GC |
| *Dricurus modestus* | Velvet-mantled Drongo | C | | F | | C | | | | |
| **CORVIDAE (1)** | | | | | | | | | | |
| *Corvus albus* | Pied Crow | | | | | x | | | | |
| **STURNIDAE (4)** | | | | | | | | | | |
| *Poeoptera lugubris* | Narrow-tailed Starling | | | | | U | | | | GC |
| *Onychognathus fulgidus* | Forest Chestnut-winged Starling | | | | | U | | | | GC |
| *Lamprotornis cupreocauda* | Copper-tailed Glossy Starling | F | b | x | | U | | NT | RR | GC |
| *Lamprotornis splendidus* | Splendid Glossy Starling | F | | U | | F | B | | | |
| **PLOCEIDAE (10)** | | | | | | | | | | |
| *Ploceus cucullatus* | Village Weaver | | | | | x | | | | |
| *Ploceus tricolor* | Yellow-mantled Weaver | F | | U | | C | B | | | GC |
| *Ploceus albinucha* | Maxwell's Black Weaver | U | | x | | U | | | | GC |
| *Ploceus preussi* | Preuss's Golden-backed Weaver | | | x | | R | | | | GC |
| *Malimbus nitens* | Blue-billed Malimbe | F | | U | | R | | | | GC |
| *Malimbus malimbicus* | Crested Malimbe | F | | R | | U | | | | GC |
| *Malimbus cassini/ibadanensis* | Cassin's/Ibadan Malimbe | | | x | | | | | | GC |
| *Malimbus scutatus* | Red-vented Malimbe | U | | U | | | | | | GC |
| *Malimbus rubricollis* | Red-headed Malimbe | F | | F | | C | B | | | GC |
| *Quelea erythrops* | Red-headed Quelea | | | | | R | | | | |
| **ESTRILDIDAE (7)** | | | | | | | | | | |
| *Parmoptila rubrifrons* | Red-fronted Antpecker | R | | x | | | | | | GC |
| *Nigrita canicapilla* | Grey-crowned Negrofinch | C | b | C | | C | | | | |
| *Nigrita bicolor* | Chestnut-breasted Negrofinch | R | | R | | R | | | | GC |
| *Nigrita fusconota* | White-breasted Negrofinch | R | | x | | | | | | GC |
| *Spermophaga haematina* | Western Bluebill | x | | | | U | | | | GC |
| *Lonchura cucullata* | Bronze Mannikin | | | x | | | | | | |
| *Lonchura bicolor* | Black-and-white Mannikin | | | x | | U | | | | |
| | **Forest biome species (this study)** | 96 | | 82 | | 97 | | | | |
| | **Total forest biome species** | 111 | | 114 | | 102 | | | | |
| | **Species per site (this study)** | 126 | | 109 | | 138 | | | | |
| | **Total species per site** | 142 | | 153 | | 146 | | | | |

## Abundance Ratings

*Draw River*

C = Common, recorded on 5-6 days
F = Fairly common, recorded on 3-4 days only
U = Uncommon, 2-3 days only
R = Rare, 1-2 days only

*Boi-Tano/Tano-Nimiri*

C = Common, recorded on 4 days or in large numbers
F = Fairly common, recorded on 3 days only
U = Uncommon, 2 days only
R = Rare, 1 day only

*Krokosua*

C = Common, recorded on 6-7 days
F = Fairly common, recorded on 4-5 days only
U = Uncommon, 2-3 days only or on 4 in small numbers
R = Rare, 1-2 days only

x = recorded during previous surveys but not in this study

## Threat Status

VU = Vulnerable
NT = Near Threatened
DD = Data Deficient

## Restr. Range

RR = Species with globally restricted range, occurring in Upper Guinea forests Endemic Bird Area

## Biome

Biome-restricted species:
GC = Guinea-Congo Forests biome

## Breeding

B = young or occupied nest observed
b = bird carrying nesting material or inspecting nest hole

# Appendix 9

List of species and numbers of
birds mist-netted in Draw River,
Boi-Tano and Krokosua Hills forest
reserves

*Hugo Rainey and Augustus Asamoah*

| Species | Sites | | |
|---|---|---|---|
| | **Draw River** | **Boi-Tano** | **Krokosua Hills** |
| *Turtur tympanistria* | | | 1 |
| *Cercococcyx mechowi* | | 1 | |
| *Alcedo leucogaster* | 2 | 1 | |
| *Indicator maculatus* | | 1 | |
| *Smithornis rufolateralis* | | 1 | |
| *Andropadus virens* | | 2 | |
| *Andropadus latirostris* | 12 | 17 | 9 |
| *Phyllastrephus icterinus* | 2 | 5 | |
| *Phyllastrephus albigularis* | | | 7 |
| *Bleda syndactyla* | | 2 | 3 |
| *Bleda eximia* | 1 | 1 | |
| *Bleda canicapilla* | | 6 | 8 |
| *Criniger barbatus* | 1 | 2 | |
| *Stiphrornis erythrothorax* | | 2 | 3 |
| *Alethe diademata* | 1 | | 1 |
| *Neocossyphus poensis* | | 4 | |
| *Camaroptera superciliaris* | | | 1 |
| *Camaroptera chloronota* | | | 1 |
| *Hylia prasina* | | 2 | 4 |
| *Terpsiphone rufiventer* | 2 | 1 | 3 |
| *Illadopsis rufipennis* | | 9 | |
| *Illadopsis cleaveri* | | 2 | |
| *Illadopsis rufescens* | | 3 | |
| *Cyanomitra cyanolaema* | | | 1 |
| *Cyanomitra obscura* | 10 | 11 | 12 |
| *Spermophaga haematina* | | | 2 |
| **Species trapped at each site** | **8** | **19** | **14** |
| **Numbers trapped at each site** | **31** | **73** | **56** |
| **Total birds trapped** | **160** | | |

# Appendix 10

## Names and museum location of voucher specimens from the 2003 RAP Survey of the Western Region of Ghana

*Jan Decher, James Oppong and Jakob Fahr*

(Museum acronyms: USNM = United States National Museum, Washington D.C., USA; ZFMK = Zoologisches Forschungsinstitut und Museum Alexander Koenig, Bonn, Germany)

| Species | Locality | Museum Catalog Numbers |
|---|---|---|
| *Crocidura buettikoferi* | Draw River | ZFMK 2003.1026 |
| *Crocidura foxi* | Draw River | ZFMK 2003.1023, -29, -30 |
| *Crocidura crossei / jouvenetae* | Boi-Tano<br>Krokosua Hills | ZFMK 2003.1039<br>ZFMK 2003.1054 |
| *Crocidura muricauda* | Draw River<br>Boi-Tano | ZFMK 2003.1031<br>ZFMK 2003.1040 |
| *Crocidura obscurior* | Draw River<br>Boi-Tano<br>Krokosua Hills | ZFMK 2003.1021<br>ZFMK 2003.1034<br>ZFMK 2003.1051, -53, -55, -56 |
| *Crocidura olivieri* | Krokosua Hills | ZFMK 2003.1057 |
| *Epomops buettikoferi* | Boi-Tano | ZFMK 2003.1037 |
| *Nanonycteris veldkampii* | Boi-Tano<br>Krokosua Hills | ZFMK 2003.1033<br>ZFMK 2003.1046 |
| *Myonycteris torquata* | Draw River | ZFMK 2003.1022 |
| *Hipposideros beatus* | Boi-Tano | ZFMK 2003.1036 |
| *Hipposideros gigas* | Boi-Tano<br>Krokosua Hills | ZFMK 2003.1041<br>ZFMK 2003.1050 |
| *Hipposideros fuliginosus* | Boi-Tano<br>Krokosua Hills | ZFMK 2003.1035<br>ZFMK 2003.1042 |
| *Hipposideros ruber* | Draw River<br>Krokosua Hills | ZFMK 2003.1024, -25, -28<br>ZFMK 2003.1045, -47 |
| *Nycteris arge* | Boi-Tano | ZFMK 2003.1032 |
| *Glauconycteris poensis* | Draw River | ZFMK 2003.1027 |
| *Neoromicia africanus* | Krokosua Hills | ZFMK 2003.1043 |
| *Pipistrellus nanulus* | Krokosua Hills | ZFMK 2003.1048 |
| *Scotophilus nucella* | Krokosua Hills | ZFMK 2003.1044 |
| *Scotophilus nux* | Krokosua Hills | ZFMK 2003.1049, -58 |
| *Cricetomys emini* | Draw River<br>Boi-Tano | USNM 589671<br>USNM 589683 |

| Species | Locality | Museum Catalog Numbers |
|---|---|---|
| *Dephomys defua* | Draw River<br>Boi-Tano | USNM 589672<br>USNM 589590 |
| *Hybomys trivirgatus* | Draw River<br>Boi-Tano | USNM 589673<br>USNM 589684 |
| *Hylomyscus alleni* | Draw River | USNM 589674, -75, -76, -77, -78 |
| *Lophuromys sikapusi* | Draw River<br>Boi-Tano | USNM 589679, -80<br>USNM 589685, -86, -87 |
| *Malacomys cansdalei* | Draw River | USNM 589681 |
| *Malacomys edwardsi* | Boi-Tano<br>Krokosua | USNM 589688<br>USNM 589691 |
| *Praomys tullbergi* | Draw River<br>Boi-Tano<br>Krokosua | USNM 589682<br>USNM 589689<br>USNM 589692,-93 |
| *Graphiurus nagtglasii* | Krokosua | USNM 589694 |
| *Grammomys rutilans* | Boi-Tano | USNM 589695 |

# Appendix 11

## Large mammals whose presence was confirmed in Draw River, Boi-Tano, Tano Nimiri and Krokosua Hills in 2003 compared to Bourlière, 1963, for Tano Nimiri

*Abdulai Barrie and Oscar Aalangdong*

Abbreviations: H = heard, S = seen, P = photographed, D = dung, sm = smoked, F = fur, T = tracks, and numbers = counts. Species in bold recorded from one site only. Scientific names are based on Kingdon (1997).

| Order | Family | Species | Common name | Draw River | Boi-Tano | Krokosua Hills | Tano Nimiri 1954 (Bourlière 1963) | Threat status |
|---|---|---|---|---|---|---|---|---|
| Rodentia | Sciuridae | *Funisciurus pyrropus* | Fire-footed rope squirrel | 2 | 3 | 4 | | |
| | | *Paraxerus poensis* | Green squirrel | 7 | 5 | 6 | | |
| | | *Heliosciurus rufobrachium* | Red-legged sun squirrel | | 3 | 2 | | |
| | Anumaluridae | *Anomalurus beecrofti* | Beecroft's anomalure | T | | 1 | | |
| | Hystricidae | *Atherurus africanus* | Brush-tailed porcupine | 3 | T | 1 | | |
| | Hystricidae | *Hystrix cristata* | Crested porcupine | | | | 38 | |
| | | *Cricetomys eminii* | Giant pouched rat | 4 | 1 | 1, P(1) | | |
| Carnivora | Herpestidae | *Herpestes ichneumon* | Egyptian mongoose | 1 | 1 | | | |
| | | *Herpestes sanguinea* | Slender mongoose | 4 | 5 | 3 | | |
| | | *Crossarchus obscurus* | Cusimanse | 2 | | 1 | | |
| | Viverridae | *Civettictis civetta* | African civet | T | T | | | |
| | | *Nandinia binotata* | **African palm civet** | H | | | | |
| | | *Manis spp.* | Pangolins | | | | 48 | |
| Hyracoidea | Procaviidae | *Dendrohyrax dorsalis* | Western Tree hyrax | H | H | H | | |
| **Proboscidea** | **Elephantidae** | *Loxodonta africana cyclotis* | **Forest elephant** | T | | | | VU |
| **Artiodactyla** | **Suidae** | *Potamochoerus porcus* | **Red river hog** | T | | | | |
| | | *Hylochoerus meinertzhageni* | **Giant forest hog** | T | | | | |
| | Bovidae | *Tragelaphus scriptus* | Bushbuck | T&D | T | D | | |
| | Antelopinae | *Cephalophus maxwelli* | Maxwell's duiker | T&D | T | 1 | 79 | NT |
| | | *Cephalophus niger* | Black duiker | 1 | T | T | | NT |
| | | *Cephalophus dorsalis* | Bay duiker | 1&2sm | 1 | 1 | 38 | NT |
| | Neotragini | *Neotragus pygmaeus* | Royal antelope | T | T&D | F | 7 | NT |
| *TOTAL* | | | | **19** | **14** | **14** | **5** | |

## South America

* Bolivia: Alto Madidi Region. Parker, T.A. III and B. Bailey (eds.). 1991. A Biological Assessment of the Alto Madidi Region and Adjacent Areas of Northwest Bolivia May 18 - June 15, 1990. RAP Working Papers 1. Conservation International, Washington, DC.

* Bolivia: Lowland Dry Forests of Santa Cruz. Parker, T.A. III, R.B. Foster, L.H. Emmons and B. Bailey (eds.). 1993. The Lowland Dry Forests of Santa Cruz, Bolivia: A Global Conservation Priority. RAP Working Papers 4. Conservation International, Washington, DC.

† Bolivia/Perú: Pando, Alto Madidi/Pampas del Heath. Montambault, J.R. (ed.). 2002. Informes de las evaluaciones biológicas de Pampas del Heath, Perú, Alto Madidi, Bolivia, y Pando, Bolivia. RAP Bulletin of Biological Assessment 24. Conservation International, Washington, DC.

* Bolivia: South Central Chuquisaca Schulenberg, T.S. and K. Awbrey (eds.). 1997. A Rapid Assessment of the Humid Forests of South Central Chuquisaca, Bolivia. RAP Working Papers 8. Conservation International, Washington, DC.

* Bolivia: Noel Kempff Mercado National Park. Killeen, T.J. and T.S. Schulenberg (eds.). 1998. A biological assessment of Parque Nacional Noel Kempff Mercado, Bolivia. RAP Working Papers 10. Conservation International, Washington, DC.

* Bolivia: Río Orthon Basin, Pando. Chernoff, B. and P.W. Willink (eds.). 1999. A Biological Assessment of Aquatic Ecosystems of the Upper Río Orthon Basin, Pando, Bolivia. RAP Bulletin of Biological Assessment 15. Conservation International, Washington, DC.

§ Brazil: Rio Negro and Headwaters. Willink, P.W., B. Chernoff, L.E. Alonso, J.R. Montambault and R. Lourival (eds.). 2000. A Biological Assessment of the Aquatic Ecosystems of the Pantanal, Mato Grosso do Sul, Brasil. RAP Bulletin of Biological Assessment 18. Conservation International, Washington, DC.

§ Ecuador: Cordillera de la Costa. Parker, T.A. III and J.L. Carr (eds.). 1992. Status of Forest Remnants in the Cordillera de la Costa and Adjacent Areas of Southwestern Ecuador. RAP Working Papers 2. Conservation International, Washington, DC.

* Ecuador/Perú: Cordillera del Condor. Schulenberg, T.S. and K. Awbrey (eds.). 1997. The Cordillera del Condor of Ecuador and Peru: A Biological Assessment. RAP Working Papers 7. Conservation International, Washington, DC.

* Ecuador/Perú: Pastaza River Basin. Willink, P.W., B. Chernoff and J. McCullough (eds.). 2005. A Rapid Biological Assessment of the Aquatic Ecosystems of the Pastaza River Basin, Ecuador and Perú. RAP Bulletin of Biological Assessment 33. Conservation International, Washington, DC.

§ Guyana: Kanuku Mountain Region. Parker, T.A. III and A.B. Forsyth (eds.). 1993. A Biological Assessment of the Kanuku Mountain Region of Southwestern Guyana. RAP Working Papers 5. Conservation International, Washington, DC.

* Guyana: Eastern Kanuku Mountains. Montambault, J.R. and O. Missa (eds.). 2002. A Biodiversity Assessment of the Eastern Kanuku Mountains, Lower Kwitaro River, Guyana. RAP Bulletin of Biological Assessment 26. Conservation International, Washington, DC.

* Paraguay: Río Paraguay Basin. Chernoff, B., P.W. Willink and J. R. Montambault (eds.). 2001. A biological assessment of the Río Paraguay Basin, Alto Paraguay, Paraguay. RAP Bulletin of Biological Assessment 19. Conservation International, Washington, DC.

* Perú: Tambopata-Candamo Reserved Zone. Foster, R.B., J.L. Carr and A.B. Forsyth (eds.). 1994. The Tambopata-Candamo Reserved Zone of southeastern Perú: A Biological Assessment. RAP Working Papers 6. Conservation International, Washington, DC.

* Perú: Cordillera de Vilcabamba. Alonso, L.E., A. Alonso, T. S. Schulenberg and F. Dallmeier (eds.). 2001. Biological and Social Assessments of the Cordillera de Vilcabamba, Peru. RAP Working Papers 12 and SI/MAB Series 6. Conservation International, Washington, DC.

* Venezuela: Caura River Basin. Chernoff, B., A. Machado-Allison, K. Riseng and J.R. Montambault (eds.). 2003. A Biological Assessment of the Aquatic Ecosystems of the Caura River Basin, Bolívar State, Venezuela. RAP Bulletin of Biological Assessment 28. Conservation International, Washington, DC.

* Venezuela: Orinoco Delta and Gulf of Paria. Lasso, C.A., L.E. Alonso, A.L. Flores and G. Love (eds.). 2004. Rapid assessment of the biodiversity and social aspects of the aquatic ecosystems of the Orinoco Delta and the Gulf of Paria, Venezuela. RAP Bulletin of Biological Assessment 37. Conservation International, Washington, DC.

## Central America

§ Belize: Columbia River Forest Reserve. Parker, T.A. III. (ed.). 1993. A Biological Assessment of the Columbia River Forest Reserve, Toledo District, Belize. RAP Working Papers 3. Conservation International, Washington, DC.

* Guatemala: Laguna del Tigre National Park. Bestelmeyer, B. and L.E. Alonso (eds.). 2000. A Biological Assessment of Laguna del Tigre National Park, Petén, Guatemala. RAP Bulletin of Biological Assessment 16. Conservation International, Washington, DC.

## Asia-Pacific

* Indonesia: Wapoga River Area. Mack, A.L. and L.E. Alonso (eds.). 2000. A Biological Assessment of the Wapoga River Area of Northwestern Irian Jaya, Indonesia. RAP Bulletin of Biological Assessment 14. Conservation International, Washington, DC.

* Indonesia: Togean and Banggai Islands. Allen, G.R., and S.A. McKenna (eds.). 2001. A Marine Rapid Assessment of the Togean and Banggai Islands, Sulawesi, Indonesia. RAP Bulletin of Biological Assessment 20. Conservation International, Washington, DC.